D1092764

THE
LANGUAGE
OF
YOUR CAT

THE
LANGUAGE
OF
YOUR CAT

MARSHALL CAVENDISH
LONDON & NEW YORK

A QUARTO BOOK

© 1977 Quarto Limited
Produced by Quarto Publishing Limited, New Burlington Street, London W.1
Phototypeset in England by Filmtype Services Limited, Scarborough, Yorkshire.
Printed by Istituto Italiano d'Arte Grafiche, Bergamo.
Print production by Mohn-Gordon Limited, London.

ISBN 0 85685 320 8

**Published by Marshall Cavendish Books Limited,
58 Old Compton Street, London W1V 5PA.**

**Text by Frank Manolson. Original photography by Michael Busselle. Original
illustrations by Andrzej Bielecki and George Thompson. Design by Andrzej Bielecki
and Heather Jackson. Editor, Philippa Algeo. Additional text by David Hardy.
Additional photography by Jon Wyand and Trevor Wood** *(full picture-credits after index)*
Art Director, Robert Morley. Editorial Director, Michael Jackson.

Contents

The communicative cat

Cats are endlessly fascinating creatures. They have occupied a special place in the affairs of man for thousands of years.

They have been worshipped and feared, loved and loathed – but never ignored.

The people with whom they share a home are totally committed in their favour. Cat-lovers will swear that their animal companions speak to them. They are right, although this rapport is obviously not in the sense that man and cat can exchange a few words on the subject of the weather. Cats are far more subtle than that.

The feline face, often so inscrutable, is capable of great expression. Indeed, almost every thread of the cat's being has a language of its own, as communicative as mere words.

Cats have a range of sounds, gestures, rituals, and seemingly inflexible habits, which all blend into a subtle and complex pattern of communication.

This may not be communication among equals, but it is certainly a dialogue between partners.

No sensible human would ever consider for a moment that he actually owns a cat. He and the animal may live in harmony under the same roof, and the man may operate the can-opener at dinner time, but the cat will never look upon him as master. The cat does not really need man at all, although humankind certainly makes feline life a great deal easier.

Beneath the thin veneer of domestication, this animal is wild at heart, with an independence of spirit that can never really be tamed. Unlike the totally dependent dog, cats can fend for themselves if they have to. Indeed, millions of alley cats do not know any other way of life.

The feline mystique

Cats have always occupied a special, if menacing, place in the world's folklore and mythology, and many strange beliefs persist to the present day.

There is no real explanation of the cat's occult reputation, but it is not unreasonable to assume that its private and slightly sinister nature, its apparently knowing gaze and traditionally nocturnal voyaging have done little to improve the overall public image.

It is easy to understand how the simple folk of long ago could have looked at a cat's eyes gleaming at them through the dark and imagined that they were seeing an evil force,

a witch in animal shape. In every country where they are known, cats have been supposed to wield mystical powers, enabling them to perform such useful functions as foretelling the future or controlling the weather.

It is commonly believed that the cat was exclusively the witch's animal familiar, but this is not so. In 1587 George Giffard wrote, in a discourse on sorcery, 'The witches hath their spirits. Some hath one, some hath more, as two, three, four or five, some in one likeness and some in another, as like cats, weasels, toads or mice.'

Many of these unfortunate creatures have found them-

selves being tried, convicted and then hanged or burned along with the witches with whom they were judged to have been in league.

The effects of a black cat's presence depend on where you happen to live. In the United States, Belgium, Spain and many other European countries, the dusky feline is a definite ill-omen. In South America this cat is thought to bring bad luck, bad health and even death. In Britain it is likely to be welcomed as a bringer of good fortune.

Perhaps this partly explains why Britain is probably the world's most confirmed nation of cat lovers.

Cats and witchcraft have long been associated in mythology. A witch (far left), executed in 1618, said she saw her mother's cat 'leap on her shoulder and suck her neck'. Three other witches (top right) who were involved are shown with their familiars, a kitten, mouse, owl and dog. The unfortunate cat has been depicted in an even less flattering light (centre right), as Satan tempting two feline angels. However, there is the brighter side. Cats also appear in many charming nursery rhymes, such as The Cat and The Fiddle (left) and (bottom right) in the illustrations for children's stories.

The inspiring spirit

Perhaps it is the cat's determined independence of spirit – often even its aloofness – which has made it so fascinating for so long. The Ancient Egyptians were the world's first great cat lovers. They worshipped a cat-headed goddess called Bast or Pasht, from which the word Puss probably comes. In addition to this, they made many beautiful figures and talismans representing cats, and these can be seen in museums all over the world. The Egyptian preoccupation with the cat went a great deal further than art, for to kill or eat one was a crime punishable by death,

With the spread of Christianity, cat worship declined. Although there is only one reference to cats in the Bible, the animal has played a significant and continuing role in Christian art, particularly painting.

A legend in Italy says that at the moment Mary gave birth to Jesus, a pregnant cat, living in the same manger, had a litter of kittens. This charming tale has influenced many artists, most notably Leonardo da Vinci, who included a cat and kittens as a part of many of his studies of the Madonna and Child. In 1504 Albrecht Dürer produced an

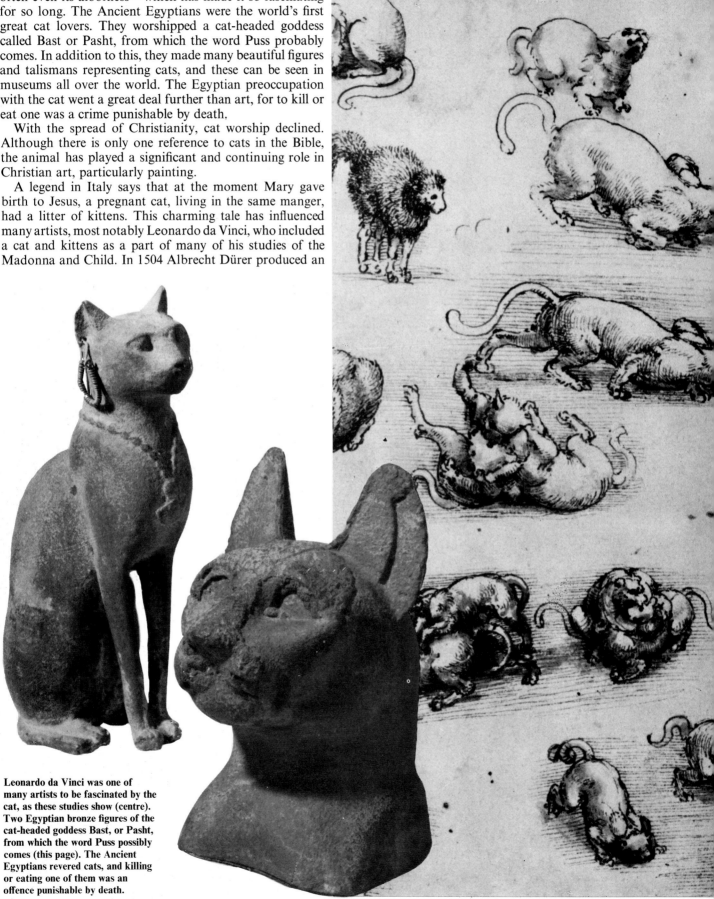

Leonardo da Vinci was one of many artists to be fascinated by the cat, as these studies show (centre). Two Egyptian bronze figures of the cat-headed goddess Bast, or Pasht, from which the word Puss possibly comes (this page). The Ancient Egyptians revered cats, and killing or eating one of them was an offence punishable by death.

engraving of the Garden of Eden, which showed a cat sitting at the foot of the Tree of Life, its tail curled around Eve's legs. The undoubted virtues of cats have continued to inspire artists throughout the centuries, from Rembrandt to Renoir, and Gainsborough to Gauguin.

Writers and poets have also come under the beguiling spell of the cat. Lewis Carroll created the vanishing Cheshire Cat – illustrated by Tenniel – for Alice in Wonderland, while Edward Lear gave the world the whimsical classic tale of The Owl and The Pussycat.

T. S. Eliot revealed himself as one of literature's great cat lovers with his Old Possum's Book of Practical Cats.

Just as enduring as any of the great works of art involving the cat, is the central role played by the animal in the development of the cartoon movie culture. Felix the Cat was the daddy of all animated felines, and these early productions can still be seen on television. They have an endearing naivety which contrasts sharply with the sophisticated sadism of Tom and Jerry, and Sylvester's equally evil attempts to kill Tweetie Pie. It's enough to make a cat laugh.

Two enduring felines. Tenniel's famous Cheshire Cat, which disappeared, leaving Alice with only a smile, was the forerunner of such 'stars' as Felix, the first animated cartoon cat. The frolics of Felix were the inspiration of the modern mayhem carried out by Tom and Jerry, and the slightly more sinister Sylvester.

Distant cousins

Cats of one sort or another have been around for 50 million years, long before man made his first fumbling appearance on earth. The first of the long line leading to the tabby curled up at the foot of your bed, was a weasel-like carnivore called *Miacis*. It had a long body and short legs and, in addition to being the predecessor of the cat, was probably the grand-daddy of the dog and the bear as well.

It took another 10 million years for the first cat-like carnivore to appear (and yet another 10 million before the dog made its bow). This first cat is known as *Dinictis* and was about the same size as the lynx. It looked very much like the modern cat. However, although its canine teeth were larger, its brain was considerably smaller.

Over the next 10 million years or so, *Dinictis* appears to have split itself into two groups. One of these included the family Felidae, to which all cats – domestic, wild, lions, tigers, etc. – belong. In addition to these, *Dinictis* also includes the civet, the genet, and the snake-killing mongoose, all of whom are cousins of the domestic cat.

The second group included the great sabre-toothed cat, which roamed the plateaux and forest regions of their prehistoric times.

The North American species, *Smilodon californicus*, was one of the most advanced sabre-tooths, and its remains have been found in great numbers at the asphalt pits at Rancho La Brea, near Los Angeles, Southern California. Many examples of prehistoric big cats have been recovered from this area. Millions of years ago there were pools of water lying on top of the tar pools. Many animals which came to drink were trapped in the tar, and their bones were preserved in the asphalt.

Also found at Rancho La Brea, although in much smaller numbers, are the remains of *Felis atrox*, the American Lion. Perhaps fewer of them were trapped because they were too intelligent to be caught in this fashion. This animal was a near relation of the European Cave Lion, which was still around in the 5th century BC. Indeed, some of them are reputed to have attacked Xerxes and his army as they invaded Macedonia.

The last of the Californian line lived about 13,000 years ago. Sabre-tooths were extinct in Europe long before this. These creatures were adapted to hunt the thick-skinned mammoths and mastodons. When these giant land mammals disappeared from the face of the earth, such highly-specialized predators as the sabre-tooth went with them. It is from Felidae, the more adaptable descendants of *Dinictis*, that the 34 species of cat known to modern man have developed. The sole exception to this is the cheetah, which belongs in a genus of its own.

These 20th century members of the family have adapted and diversified to meet the wide range of climate and environment they are likely to find. They are all carnivores and highly efficient hunters. Some prefer the solitary life on the prowl, while others – the lion is a notable example – like to live and hunt in groups.

Similarities between *Felis sylvestris*, the modern European or Scottish Wild Cat, and the modern house cat, are so obvious that it seems highly probable that the Wild Cat, now extinct outside Europe and Asia, was the last stop on the evolutionary train before arriving at the cuddly creature purring in front of the fire. The Wild Cat hunts small mammals like rabbits and hares, but because it sometimes attacks poultry it is being gradually destroyed in Europe.

The small Indian mongoose, *Herpestes auropunctatus* (left), is a descendant of the group which once included the cat.

The civet, *Vivera civetta* (left), from Africa, is another creature believed to be a relative of the cat, a relationship that goes back 30 million years.

Only size seems to separate the tiger from its smaller cousin, the house cat. They can both look fierce to an opponent.

The sleek-looking ocelot shares the vice of laziness with its domestic counterpart.

This leopard cub is seeking food, or perhaps reassurance. At this age, the only sound it can make is a kitten-like mew.

The blotched genet, *Genetta tigrina* (left), is the most cat-like of the early relations, as the picture below clearly shows.

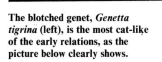

Feline climes

The earliest examples of the cat almost certainly came from Asia, Europe and North America. It was relatively simple for them to reach the African continent but it took them until much later in time to arrive in South America, because for long periods this continent was an island.

The cat could not reach Australasia until man took it with him. To this day, Australia, New Zealand and the Oceanic Islands do not have an indigenous cat. The same is true of the West Indies and Madagascar, for example.

Once the globe was colonized by man, cats began to diverge and form new species. Today there are 34 species, all members of the family, Felidae. Most of them live in the warmer climes, although the Snow Leopard is found in Tibet, and there are some lynxes in Scandinavia and Canada.

As far as cats are concerned, America can be divided into North and South, for only the puma is common to both. One could be forgiven for doubting this, for in the United States alone, the puma is known as cougar, mountain lion, painter and catamount.

Many theories exist about the origins of the domestic cat. Most experts believe that the earliest came from Egypt, and that the Kaffir, or Caffre, Cat is the daddy of all European short-hair types. Sadly, no proof exists.

Many of the world's Wild Cats, including the European or Scottish Wild Cat, will interbreed quite freely with their local domestic counterparts.

Some points of departure

The Russian Blue
Prior to 1900, this cat was known as the Archangel Blue and is thought to have been introduced into Britain by sailors trading from the Baltic port of Archangel in 1860.

The Siamese
The first recorded pair of Siamese cats were introduced into Britain by the English Consul General for Bangkok in 1884. They were exhibited at Crystal Palace Cat Show in London, in 1885 and appeared in the United States of America in 1900.

The Abyssinian
The first reliable reference to the breed was in 1868 when Lord Robert Napier returned to Britain from Abyssinia (Ethiopia) after a military expedition, bringing one of the cats with him.

The Birman
The Sacred Cat of Burma first appeared in France in 1919 when a pair of cats were sent to two soldiers who had come to the assistance of the priests of a Burmese temple. The male did not survive the trip but the female was pregnant and thus ensured the survival of the breed in the west.

The Burmese
Burmese cats may be traced back to a single brown female named Wong Mau which was introduced into the United States of America from Burma in 1930.

The Chinchilla
The original breed is credited to a Mrs Vallence who in the 1880's mated a smoke-coloured cat to a silver tabby. In the resulting litter was the first Chinchilla male called Silver Lambkin, which was exhibited at the Crystal Palace Show in 1888.

The Manx
This cat is said to originate from the Isle of Man, but cats without tails occur in other parts of the world, such as China and Russia.

The Rex
This breed was first discovered in 1950 on a farm on Bodmin Moor, Cornwall, England.

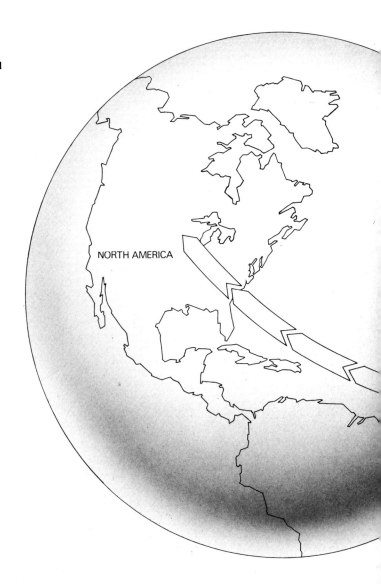

NORTH AMERICA

This Scottish Wild Cat (left) was caught as a kitten 'swimming' in a sea loch. It survived the dowsing to become a fine example of the species. It bears a close resemblance to the tabby (far left), but it is far more heavily built, and has a shorter, thicker tail and a stouter head. The Scottish Wild Cat and its European cousin are the only survivors of *Felis Sylvestris*. The Wild Cat will breed quite freely with local domesticated cats.

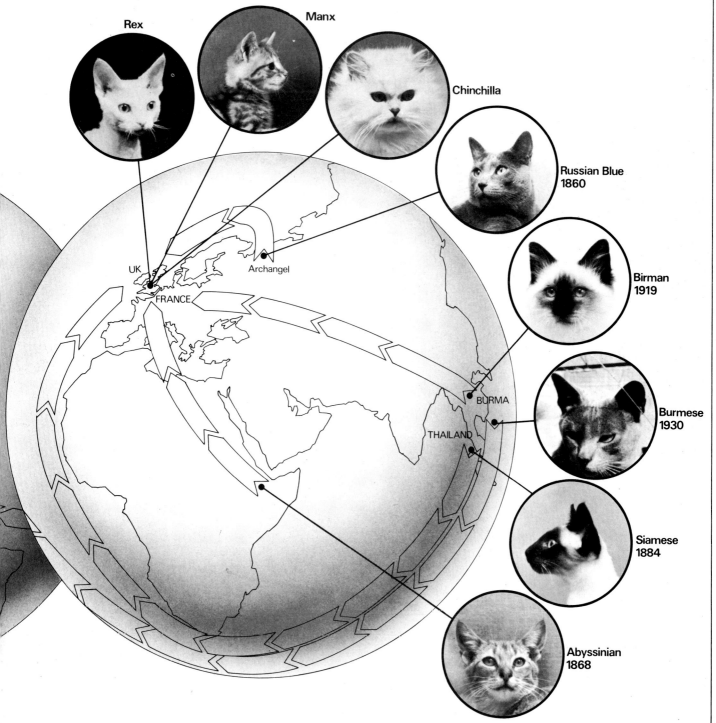

Rex

Manx

Chinchilla

Russian Blue
1860

Birman
1919

Burmese
1930

Siamese
1884

Abyssinian
1868

UK

Archangel

FRANCE

BURMA

THAILAND

Breed recognition

Foreign short-haired cats

The term foreign is not meant to indicate exotic lands of origin, although some of the ancestors of these animals probably came from faraway places. The group includes types developed in North America, Europe and Britain.

In general, the term can be said to indicate the appearance of the cat, with its slim, sophisticated shape, long tail and slender legs. The heads of these beauties are wedge-shaped with large pricked ears and oriental slanted eyes.

Most of them dislike the cat's usual solitary way of life and insist upon the company of humans, or of another cat if there are no available people.

They are the easiest of cats to groom, requiring only attention from a short-bristled brush, followed by a purr-provoking rub with a chamois leather to keep up the polished sheen.

It is important to have such foreign types as Siamese, Burmese and Russian Blues inoculated against feline enteritis, to which they are particularly susceptible.

Long-haired cats

Long-haired cats have for many years been grouped under the very generalized heading of Persian, although this term is no longer officially recognized in Britain. What is generally accepted is that these animals made their first appearance on European soil towards the end of the 16th century.

The man credited with having introduced them is the French scientist and archeologist Nicholas Fabri de Peirasc, who is believed to have brought one home with him from Turkey. The resulting long-haired varieties are considered by many people to be the most beautiful family of felines.

Varieties differ in colouring, but they must conform to certain basic standards. They should have a long, flowing coat with a fine, glossy texture, standing off from the body. The immense ruff should continue into a frill between the legs, and the tail should be full. The long-haired cat is crowned with an imposingly broad head, well-covered ears, short nose, full cheeks and a broad muzzle.

Domestic short-haired cats

Domestic short-haired cats are among nature's most intelligent and charming creations. They love to be a part of the family and are usually more than willing to share their homes with other animal inmates.

These delightful creatures are active and graceful and show great interest in everything going on in the house and garden. Definitely on the credit side, they are not so destructive to furniture and curtains as some other breeds. They tend to sharpen their claws outside, often reserving a favourite tree for the purpose. They are also great believers in exercise and this keeps them and their coats in top condition.

The short coat is easier to keep in good order, but this does not, of course, mean that it can go without regular, thorough grooming.

When buying one of these magnificent cats, you should look for a round, broad head with width between the ears, full cheeks, a short, broad nose, good muzzle, with neat small ears, rounded at the top and never broad at the base.

In general, the foreign short-hair has a slim body, long tail, and slender legs, contributing to a very sophisticated appearance. The head is wedge-shaped, with large pricked ears and slanting eyes. Some individual breeds may differ, but most foreign type cats do not like a solitary life and demand human company.

The long-haired cat should have a cobby body, set on short, thick legs. The head should be round and broad with full cheeks, a short, snub nose. The ears should be small, neat and spaced well apart, and the eyes should be large and round. The cat's tail should be short and thick, and the coat long and silky, with no woolliness.

The domestic short-haired cat is a sturdy animal, with strong bones and a thick body set on short, but well-proportioned, legs. The head is apple-shaped, with the skull rounded at the top. Its cheeks are well-developed, and its nose is short and broad. Prospective buyers should look for big, rounded eyes and small, slightly rounded ears. The coat should always be short and fine.

The Siamese

Like many other breeds, the Siamese is the subject of many legends and fables. Sadly, few of them have any factual basis. Some say the animal was developed in the last century from an Albino given to the King of Siam, crossed with a Temple cat and bred in the Royal Palace.

However, a picture of the cream and dark brown Seal Point Siamese appears in a set of pictures and verses made more than 400 years ago in the Ayudha period of Siam.

In 1794 a German explorer named Peter Pallas published a drawing of a cat which he saw while travelling around the Caspian Sea. It had a light brown body and black markings, much like the points – that is, ears, mask, legs, paws and tail – of the modern Siamese.

Although no really solid evidence exists, it is now generally accepted that the Siamese does have Eastern origins. However, there are now more of them in such places as Iowa than in Siam, or Thailand as it is now.

The Seal Point is the most popular with owners everywhere, but there are several 'colour schemes' to choose from: Blue Point, Chocolate Point, Lilac (or Frost) Point, Red Point, Tortoiseshell Point, Cream Point, Albino, and Tabby (or Lynx) Point.

Red, Tabby and Tortoiseshell types are not universally recognized and some authorities classify them as Colour-point Short-hairs.

Whatever the colour or label, Siamese remain by far the most popular pedigree cats, both in the home and at shows. It is easy to see why, for these blue-eyed creatures have a most appealing disposition, cramming unbelievable vitality into their frail-looking bodies.

They are usually very intelligent, cunning and resourceful, with an almost limitless ability to learn tricks and invent games. The owner should beware at mealtimes, for the sneaky Siamese is uncomfortably efficient at stealing the food from the plate while human heads are turned.

Siamese are the most popular of pedigree cats, both in the home and at shows. The modern example is expected to have a wedge-shaped head, with a smooth outline, narrowing to a fine muzzle. The cheeks should not be pinched, neither should they be rounded.

The Burmese

There are two categorical statements to be made about the Burmese. The first is that you would be extremely lucky to see one anywhere near Burma, and the second is that it was developed in America in 1936. Anything else said about the animal's origins is make-believe.

It did not make its European bow until after World War Two, but was an instant success, being a good deal closer than the Siamese to what cat fanciers regard as the ideal oriental type. This is because Burmese enthusiasts have maintained constant vigilance in preserving the looks that contribute to its overall attractiveness.

Burmese have Siamese-looking, long, lithe bodies, pointed heads and slanty yellow eyes, but their colour is different. Most of them are brown, although there is a popular, but rarer and more expensive, Blue Burmese.

Further variations include Chocolate, now accepted as a separate breed, and Lilac, Red Tortoiseshell, Cream, Blue-Cream, Chocolate Tortie and Lilac Tortie, which are all recognized only in Britain.

The Burmese is a most civilized creature, less likely than the Siamese to claw the furniture or shin up the curtains. Also it is less vocal and less highly strung.

They are, however, just as aggressively friendly but, unlike the Siamese, they do not fret if you pop out to the shops, leaving them on their own.

The Burmese has a definite Siamese look, with a long, lithe body, pointed head and slanted yellow eyes. This elegant animal should be of medium size, with a hard and muscular body. It should have a strong, rounded chest, and a straight back from shoulder to chest. The legs should be slender with the hind pair longer than the front, and the paws neat and oval-shaped. The medium-length tail should taper only slightly.

The Russian Blue

The Russian Blue's coat is really grey and he is unlikely to have many relatives behind the Iron Curtain, although the first may have been brought back from Russia by British sailors in Elizabethan times.

Around 1900, one was taken across the Atlantic for the first time and made its home in Chicago. The resulting cat is known as American Blue, while many others around the world know the breed as Maltese Cats.

Today's animal has a long, graceful body, a fairly long tapering tail, long legs, oval feet and a short, wedge-shaped head. Its eyes are always green and, ideally, are almond-shaped. Its coat is short and lustrous, and is easily groomed to sleek perfection.

The Russian, or American, Blue has a deserved reputation for gentleness. It can be extremely shy and has a very quiet voice. These appealing creatures become greatly attached to their owners and seem to adapt particularly well to life in apartments. This may make them the ideal feline companion for the city-dweller.

American and British standards for the Russian Blue both require a long, graceful body, long legs, small oval feet and a tapering tail. Its short, wedge-shaped head should have a flat, narrow skull and straight forehead and nose. The almond-shaped eyes should be set wide apart and be a vivid green when adult. The skin on the large pointed ears should be very fine and almost transparent.

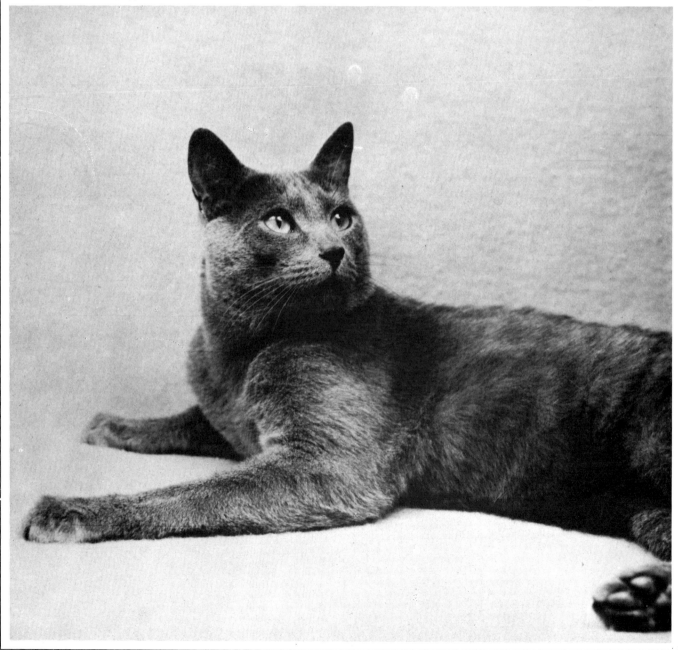

The Abyssinian

Many owners of Abyssinians like to think of their aristocratic pets as being direct descendants of the cats of the Pharaohs. Unhappily, they are almost certainly deluding themselves.

It is also said that in 1869 a Mrs Barrett-Lennard, wife of a British Army officer, came home from Abyssinia with a cat called Zulu. This is reputed to have been the first of the line. However, a portrait published six years later reveals a cat totally unlike the breed known today.

The truth about the Abyssinian, far less romantic than the legend, is probably that the creature was manufactured in Britain at the turn of the century by the skilful breeding of Tabbies.

Of course, it could be in part a reversion to the form of the ancient cat. If so, it has not inherited many of the characteristics of its Egyptian ancestors. These fiery cats were reputed to have ridden into battle on their masters' backs, springing at the enemy with claws out.

The modern Abyssinian is not a quiet cat, but it is cautious. It will always survey the territory with care before venturing a paw, making sure of escape routes before setting off. Like the Siamese, it will explore things with its paws rather than nose. If the object under examination is a lighted cigar or cigarette, this is likely to be painful.

Since 1963, the Red Abyssinian has been a recognized separate breed. It differs from its cousin only in colour, which should be a richer copper red.

The best Abyssinian specimens are bred in the United States. Prices are high, but so is the quality.

The Abyssinian has a slender body, tapering tail, and slender legs, with neat oval feet. The American standard demands a slightly blunter muzzle than Europe, and a wedge-shaped head, without flat planes. The brow, cheeks, and profile lines all show a gentle contour. The eyes should be large, and the ears sharply pointed.

The Chinchilla

One of the most admired of the long-haired breeds is the Chinchilla. It looks delicate and almost fairy-like, but do not be misled; in this case, looks are deceptive. This is a tough and durable animal, which does not need any coddling from humans and is quite happy with an outdoor life.

The breed is generally supposed to have originated as recently as the 1880s, when a Mrs Vallence, of London, mated a smoke-coloured cat to a Silver Tabby. One of the females of the resulting litter gave birth to the first Chinchilla – a handsome fellow named Silver Lambkin.

This celebrated cat was a great champion, sweeping the board at the 1888 Crystal Palace Show, in London. He died at the venerable age of 17, but his body was preserved for posterity. Cat-lovers can still feast their eyes on the first of the line. Nearly 100 years later, Silver Lambkin remains on show at London's Natural History Museum.

Chincilla kittens demonstrate their tabby ancestry, for they are born with tabby rings on their tails, and often have markings all over their coats. These disappear as they grow, making way for the characteristic pure white.

American and British standards dictate that the undercoat should be white. The fur on the back, flanks, head, ears and tail has black or silver tips, giving the animal its sparkling silver appearance. Although the legs may be slightly shaded with tipping, the chin, tufts, feet, stomach and chest should always be pure white. The broad, round head should have good breadth between ears and muzzle. The Chinchilla's eyes should be large, round and emerald green or blue-green in colour.

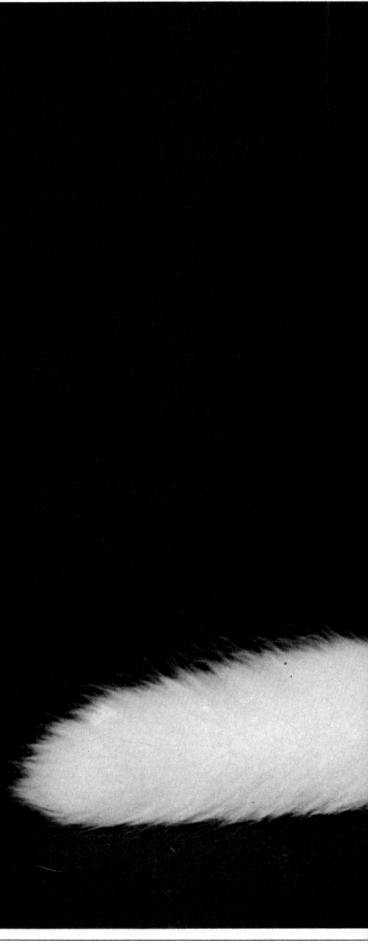

The Chinchilla is a much-admired animal and, although it looks delicate, is tough and durable. Standards in both America and Britain dictate that the undercoat should be white, while the fur on back, flanks, head and ears should have black or silver tips. The broad, round head should have breadth between muzzle and ears.

The Persians

No one knows quite where the Persian comes from. Some experts refuse to look any further than the obvious, Persia; others insist that Turkey was the starting point.

To make matters even more confusing, naturalists have advanced the slightly more exciting and spectacular theory that the Persian is descended from the Asiatic Wild Cat.

Whatever the truth, there can be little doubt about its popularity. Apart from the Siamese, the thoroughbred which appears in the greatest numbers at cat shows is the Blue Persian (or Blue Long-haired, as it is sometimes known).

Blue Persians have been popular ever since 1889 when Queen Victoria bought one, thus setting an instant trend in society. The animal's position in the social register was indelibly underlined when the Prince of Wales, later Edward VII, gave away an autographed photograph of himself to the owner of a winning Blue.

Before this the Blue had been unkindly dismissed by some authorities as merely a 'weakened' Black. These so-called experts could not have been more wrong. Today, the overall excellence of the Blue is unchallenged, and they have been used in cross-mating to help produce the high quality seen in present day Blacks, Creams and Whites. The Blue Persian should always have copper-coloured eyes.

White Persians are among nature's strangest feline creations and they can be divided into three categories:

blue-eyed, odd-eyed and orange-eyed. Sadly, those with blue eyes are almost always born deaf. However, if a kitten has even the slightest smudge of a darker eye colour, its hearing is unlikely to be affected.

The White odd-eyed is unique in having one blue and one orange eye. Breeders of the blue-eyed type are always anxious to find them, for they do not usually have hearing problems. However, some people claim that they may occasionally be deaf on the side of the blue eye.

By far the soundest of the Whites is the orange-eyed. It was developed when blue-eyed types were mated to other long-haired cats with orange eyes. It has never suffered from defective hearing.

The Persian White (left), and its Blue counterpart (right), are considered by many to be the most beautiful of cats. Persian varieties may differ in colouring, but the standards require them to have a long coat, with a ruff continuing between the front legs. The tail should be short and full, and the head should be broad, with small, well-covered ears.

The Birman

The majestic Birman is the Sacred Cat of Burma, but it should never be confused with the Burmese, to which it is not related in any way. It has one of the most colourful and attractive histories of any feline variety, having originated in the great temples of its native land, and having narrowly escaped extinction during World War Two.

Birmans are still revered in Burma, where many people believe them to be reincarnations of dead priests. A charming legend says that centuries ago, Sinh, a loyal white cat, placed its paws on the face of Khmer priest Mun-Ha as he died after being struck down by temple raiders. Sinh then turned to face the golden statue of the goddess Tsun-Kayn-Kse, who watches over the transmutation of souls with her sapphire-blue eyes.

As it did so, the cat's white body was turned to gold and its yellow eyes to blue. Its legs became brown, but the paws resting on its dead master's face remained white. Next morning the 100 other temple cats were also golden. Seven days later Sinh too died, taking Mun-Ha's soul to paradise.

The more modern history of the Birman has been more down to earth, for in 1945 only a single pair survived, and it took years for the variety to recover. Happily, this magnificent animal is now flourishing once again.

Birmans have a broad, rounded head, with medium-length, heavy ears. The body should be long, on low legs with short strong paws. Its coat should be long and silken-textured, with a heavy ruff around the neck and slightly-curled hair on the stomach. The tail should be bushy and of medium length.

The body of the beautiful Birman is long and low, set on short, strong legs, with a longish tail. The head should be wide and built strongly, with full cheeks, and slightly flat above the ears. In America, a Roman nose is also required, medium long with low set nostrils. This handsome creature's fur should be long, with a full ruff and bushy tail. The texture should be silky and slightly curled on the belly. A most notable characteristic is, of course, the Birmans 'white' toes.

The Domestic Blue and Black

As with so many other excellent breeds, the British Blue short-haired cat was almost wiped out by World War Two. To stave off the threat of extinction in 1945, foreign-type cats were used in experimental mating programmes.

This nearly added the death blow, for the essential bone-structure of the Blue began to be altered. It took another decade to put things to rights. In the mid-Fifties some breeders sent their British Blue queens to mate with selected Blue Long-hairs. Results were good and the breed restored.

It seems that the future of many British short-haired cats lies in the occasional similar crossing to preserve the type and bone structure.

In spite of its traumatic experience, the British Blue remains an essentially quiet cat, wishing for nothing more than a tranquil life. It is happy to go outside for a romp – but even happier to come indoors and lie in front of the fire.

The tranquility of this sociable animal conceals a degree of stubbornness that seems always to ensure that he owns you and not the other way round.

Its Black short-haired counterpart is probably the most maligned of all living creatures. For centuries it has been associated with witchcraft and many other mystical happenings. The black cat of ill-omen is usually thought of as a slinky creature, with more in common than the Siamese than any other. The more rotund and friendly looking British Black could surely never be anything other than the bringer of great good fortune.

Daily brushing is essential to maintain the dense and glossy coat which makes this animal stand out. This is helped by a good diet including plenty of raw beef with the occasional raw egg yolk. Apart from this, the cat itself will do the rest, for it is almost obsessed with cleanliness.

The British Blue (left) and the Black (right) are among the most popular short-haired breeds. The Blue should be level in colour, with no tabby markings, shading, or white of any sort. Its large, full eyes should be copper, yellow or orange. The Black's coat should not show rusty tinges or any trace of white. Its eyes can be of deep copper or orange. Any trace of green is regarded as a fault.

The Tabbies

Of all tabby breeds, the Silver is by far the most popular. This animal is gentle and loving, with an affecting shyness. The females are easy to spot, for they have very pretty 'smiling' faces. Silver Tabbies also manage to get on well with other breeds, and the Siamese is a particular friend.

They have been highly-regarded, and highly-priced, for more than 100 years. However, World War Two caused a great rundown in quality. In Britain, for example, only one pedigree female could be found after the war. Happily, the Forties saw a big revival and the Silver is once again in its rightful place.

The Brown Tabby type is quite common among the mongrels, but the pedigree variety is the rarest of all tabbies.

This is a pity, for the Brown Tabby kittens are a constant joy from the moment of birth. They are exceptionally playful creatures with a highly-developed sense of fun. They also have hearty appetites and a strong sense of the family bond. Browns are a pleasure to have around, for they are among the cleanest of cats and will housetrain with the greatest of ease.

As the cat grows up you must be prepared to give it lots of your attention, for it is the complete extrovert and will go to almost any length to be the centre of attraction.

In these liberated days it is interesting to note that the female is the most aggressive, tending to dominate the less-argumentative male.

The Red Tabby has an unfortunate public image that it finds difficult to live down. It is often, incorrectly, dismissed as 'that ginger tom from next door', or as a 'marmalade' cat. The properly-bred Red Tabby is definitely neither ginger nor orange, and perhaps the most beautiful of the tabby types.

Its base coat should be a rich red, with markings of an even deeper tone. Correctly, these should be three dark stripes down the back, with an oyster-shape on the sides and a butterfly mark on the back of the neck. Clear markings should encircle the neck, throat and legs, while the tail is ringed all the way down to the tip, which should be of the darker colour.

In many areas the word tabby is believed to mean female, but this too is wrong. The word almost certainly comes from the markings on the watered silk made at Attabiy in ancient Baghdad.

A conflicting tradition in many country areas says that all Red Tabbies are male. This is not true either, but it is a fact that red-to-red matings produce twice as many males.

Early books on cats contain few references to the Red Tabby. A book called The Domestic Cat, published in 1876, says of them, 'They are the prettiest of pets and the honestest of all cat kind. They are such good ratters that neither mice nor rats will frequent the house they inhabit.'

For all its obvious attractions, the Red Tabby has never managed to achieve the popularity of its cousins.

Whoever first used the expression 'the cat's pajamas' was undoubtedly thinking of the tabby, whether Red (below), Silver (right), or any of the other colourings. The standard tabby pattern is strictly laid down, and is not possessed by all those pet cats commonly called tabbies. The markings should always present a clear contrast and be free of all brindling.

The bi-coloured breeds

Bi-coloured cats, those with two colours in their coats, have been known for many years, although standards have altered over the last century. In the early days as many colours as possible were encouraged: black and white, white and black, blue and white, and tabby and white animals of every hue.

However, in 1971 the standard was revised and now the recognized combination is any solid colour and white. Patches of colour must be clear and evenly distributed. Not more than two-thirds of the coat should be coloured, with a maximum of one half to be white. The face should be patched, ideally with a white blaze. Any tabby markings are now regarded as a definite fault. They are exhibited in the 'Any Other Colours' class, and are sometimes known as 'magpie' cats, because the most common combination is black and white.

From the start, black and white cats have been the most popular of this breed. Ideally, the coat should be a dense brown-black, marked with white. As with modern Birmans, the feet should be white, with white on the chest and face.

In 1876, Dr Gordon Stables wrote of them, 'A good black and white cat is a very noble looking animal. If well-trained and looked after, you can hardly have a nicer parlour pet.' However, he added, 'If well-treated black and white cats are apt to turn a little indolent and lazy.' The mixed ancestry of these cats might be responsible for their good health, longevity and easy breeding. They grow quite large and are very hardy.

The bi-coloured cat, whether black and white, ginger and white, or patched all over, is expected to conform to standards. There should be no tabby shading in the dark-coloured portion. Markings should start behind the shoulders, and include the tail and hind legs, leaving the hind feet white. The mask should be divided in half, and eyes copper, orange or amber.

The Cream

Cream short-hairs are notoriously difficult to breed and, because of this, are still comparatively rare. Occasionally, a natural cream is born, but it is more usual for them to be striped or barred. There is, however, very little contrast in colour and few really good specimens are seen at today's shows.

The Cream is presumed to be a development of the Red Tabby because, genetically, cream is a dilute of red.

This variety has been recognised in the United States only in the last 15 years, and is still not known in the British Isles. It is accepted in America in its 'classic' or 'mackerel' form.

This domestic short-haired pedigree animal must be a rich cream colour, without any bars, and free from white hairs.

The Manx

The origins of the mysterious short-haired cat known as the Manx are unknown and its name is wreathed in legend. The Phoenicians are said to have brought back tailless cats from trading forays to Japan. Their purpose was not the high-minded one of establishing a breed, but merely to keep down shipboard rats and mice.

The Manx is, of course, instantly identified by its complete lack of a tail. Although there are other bob-tailed cats, the good example of Manx actually has a little hollow where its tail would be. Anatomically, it has only three tail bones instead of the usual 19 or 21.

One charming legend says that a wise old Manx mother cat saw the bushy tails of cats being cut off to decorate the helmets of warlike soldiers. To save her kittens from suffering this indignity, she bit them off herself. When her offspring were grown up they were instructed to do the same to their kittens, and so it went on until we find today's familiar tailless cat.

This tale of the tail comes from the Isle of Man itself, which is more than can be said for the cat.

It is believed that these animals first arrived on the island in 1588 when a Spanish Armada ship was wrecked off the coast. Two or three intrepid cats are supposed to have braved the crashing waves to swim ashore and establish a colony in the land whose name they now bear.

The Manx is an inquisitive creature, intelligent and faithful and at the same time courageous and independent. It has long hind legs and a short back, so that it seems to bob along like a rabbit. Indeed, some early observers of the breed believed it to be part rabbit. This strange gait does not seem to impede the Manx in any way, for it is a very swift mover, nor does it lack any sense of balance.

The Manx cat is, of course, instantly recognizable by its lack of tail. A good example of this type will even have a little hollow where the tail would be. Other cats have up to 21 bones in their tails – the Manx has only three, and they are not visible. This cat should also have long hind legs and a short back, so that it actually seems to bob like a rabbit when running.

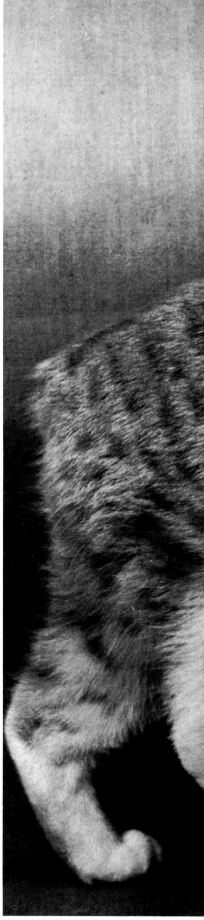

The Rex

Until 1950 there were only two types of hair on cats: long and short. Then, on Bodmin Moor, Cornwall, in England a curly-coated kitten was born. The mother was a tortoiseshell and white farm cat, the father was unknown.

Fortunately, the kitten's owner had bred and shown Astrex rabbits and she recognised the characteristic type of coat.

A geneticist was called in and he suggested that a new genetic mutation – a natural alteration of the pattern of a pair of genes – had taken place. He advised mating the kitten back to the mother when it was old enough.

After a series of test matings it was proved that a new mutation had taken place, a rare event in feline history. It was decided to name the new breed Cornish Rex, after its county of origin, and the rabbit whose coat it resembles.

The original kitten died while still young, but his son became a stud. This in itself was unusual, for it was a blue-cream and cats of this pattern are almost always female or sterile. This one, however, successfully sired several litters before reverting to type and becoming sterile.

If Cornwall leads, then neighbouring Devon will not be far behind, and so it was that in 1960 the first Devon Rex was born to a tortoiseshell and white stray, befriended by Miss Beryl Cox, who lived near a disused tin mine at Buckfastleigh, Devon.

It was known in the locality that an elusive large cat with masses of curls lived in the mine and it was assumed that he was the father. The kitten was named Kirlee.

Experts believed that he was just another close relative of the Cornish version. Breeding proved them all wrong, for when mated with a Cornish, only plain-coated kittens were born. It became plain that Kirlee was genetically different.

He certainly looked different, with mole-grey colouring and a wide-cheeked pixie face, short nose and batwing ears. The Cornish Rex had a much longer Roman nose, and this proved to be the chief difference between the two animals.

Both types are now firmly established, with offshoots in East Germany and North America.

The Cornish Rex (below left) has shorter, thinner hair than most cats, and its coat should curl, wave or ripple. Whiskers and eyebrows should be long and crinkled. Its body should be hard and muscled, with long straight legs and small oval paws. The Devon Rex (right) is completely different. It has a wide-cheeked pixie face, short nose and imposing batwing ears. The British standard insists on a broad chest and slender neck. The Devon Rex is not universally recognized in the United States.

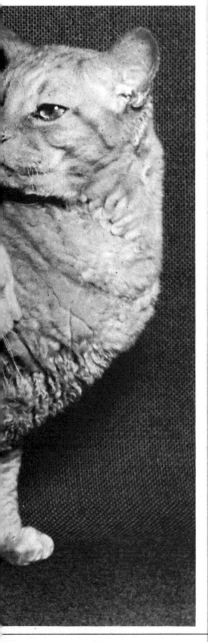

The Mongrels

The overwhelming majority of the world's cat population falls into the category affectionately known as the alley cat or moggie: an all-sorts mongrel mixture of breeds, brought about by centuries of enthusiastic promiscuity. The mongrel is most often short-haired, although no strict rules can ever apply, and can be a walking riot of all the recognized colours.

The competitive world of the show, with its thousands of dollars and pounds at stake in prize money, is not for them. They are content to be loved for their own sake. What they lack in aristocratic ancestry they make up for in charm. Black, white, blotched, patched and striped, they do not conform to any standards. Variations on the non-pedigree tabby theme are almost endless, although they never quite match the richness of their well-bred cousins. The tabby mongrel comes in silver and black tiger-stripes, or fawn and black, or brown. Their coats often show gleaming smart white 'shirts' and 'socks'.

The tortoiseshell alley cat shares some characteristics with its pedigree counterpart. It too is almost invariably

The mongrel, or moggie, is the same the world over. Whether it lives in cosmopolitan Paris (far left), the precincts of a Roman art gallery (below) or in and around London, this all-sorts mixture of type and personality is a walking riot of all the recognized colours.

female, with the rare male inevitably sterile. It can show a lovely mixture of ginger, black and white, plain or with stripes, or black mottled with ginger.

All these cats should be 'doctored' or neutered, and this particularly applies to the tom. If he is not treated, he will cause a thoroughly anti-social smell in the house and will go off for days on end in search of willing lady companions. At such times he is quite likely to make a noisy nuisance of himself and will drive your irate neighbours to throw things at him. The unfortunate tom will also get into fights and, of course, runs a higher risk of being hit by a car.

The untreated female will produce an endless succession of litters. Owners may be delighted with the first or second happy event, but after this they become a problem. Even a family with a wide circle of friends will soon run out of homes for the unwanted kittens.

Males can be castrated at any age up to five years, but it is preferable to wait until the animal is nine months old. It is never too late to spay the female, which can safely be treated after the first six months.

A perfect machine

The cat is an active and agile mammal. Its whole shape has evolved over the centuries to allow the rapid movements that we know so well. The sudden spring, the pounce, the impressively athletic standing jump, the seemingly effortless climb and the devastatingly fast short sprints are all outstanding characteristics of the cat family.

An alert head, carried by a deceptively powerful neck is specifically designed by nature to aid the cruelly graceful search, location, kill and removal of the prey, whether it is a mouse or bird in the back garden, or the lion's more substantial target in the plains of Africa.

The cat's chest is so powerful that one can hardly see the breathing movements while the animal is at rest. However, some elderly cats, or younger ones suffering from conditions such as pneumonia, or pleurisy, may find it painful to breathe using their chest muscles. They switch to use the abdominal muscles. It is common to many species in trouble and it is one of the signs which the vet should be able to spot from a distance.

Some experienced vets can even sit on a corral fence and signal to the cowhands those steers they want cutting out of the herd for further attention.

The same efficient breathing mechanism also helps the cat to communicate. The air passing across the vocal cords in the throat produces the purring sound with which we are all so familiar. Generally, we associate this feline sound with pleasure, or a not very subtle form of begging. It is less commonly realized that the purr may also indicate extreme

discomfort – or, in the aging cat, impending death. The forced breathing, combined with weakened vocal cords, produces the same sounds one would expect to hear from a perfectly healthy and happy cat.

The healthy cat's forelegs are well-adapted to climbing, sprinting, holding, or merely washing. This is because the long forearm bone, the radius, can move around its accompanying bone, the ulna, while the wristbones remain in a fixed position. At the same time, the shoulder is able to move very freely.

The hind limbs can move only backward or forward, but they are propelled by very strong muscles. Thus the cat which appears to be completely in repose can suddenly lift off with greater efficiency than the best jump jet.

Alert, aware . . . and, if you are a small rodent or bird, deadly. The cat in its various shapes and sizes is nature's design marvel, a true killing machine. It has evolved over centuries into an athletic mechanism to search out and catch prey almost in one easy movement. This animal has so many attributes at which we can only marvel: the spring, pounce and standing jump. The structure of its legs is particularly well designed and adapted to climbing, sprinting or the mundane, but essential, business of grooming.

The flexible feline

The skeleton of the cat is a magnificent example of physical engineering. Unlike the dog's skeleton, which has a wide range of variations, the cat's original shape has been retained with only slight changes.

The three functions of a skeleton are to dictate the shape, to protect the vital organs, and to provide a framework to which the body's other parts are attached.

Where the cat is concerned, there are about 230 bones, varying in size and shape. These are held together by more than 500 muscles. By far the strongest of these muscles are in the lumbar region and the hind legs, and in the neck and shoulders. The whole structure is covered by a skin which is designed to hang loose, thus giving the animal another aid to agility.

The elastic skeleton . . .

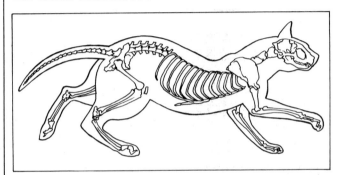
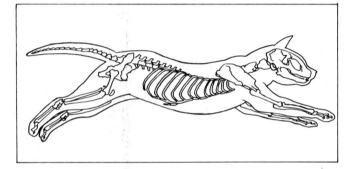

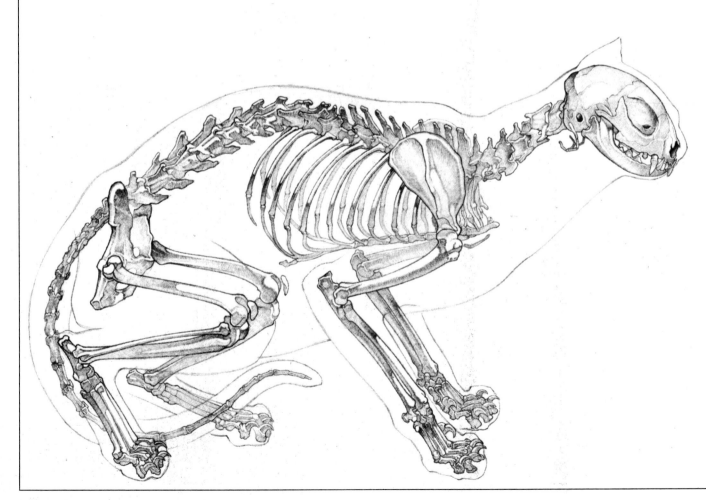

Far left: a cat's eye view with an almost unbelievable difference. This is an excellent example of the power and flexibility of the cat's neck and shoulder muscles. Here they enable the animal's head to perform a complete about turn. All this to gaze at the ceiling! The cat can get into many strange positions. It can also get out again (left). Virtually nowhere in the home is off-limits to the curious feline and, with the aid of a supple and almost elasticated spinal column, getting out from under a cupboard or a dresser is simplicity itself.

. . . and vigorous muscles.

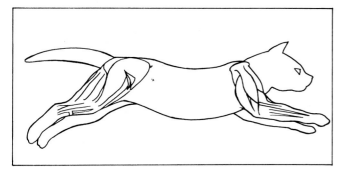 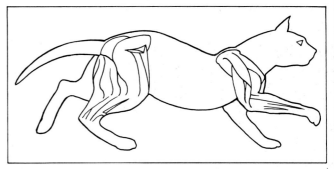

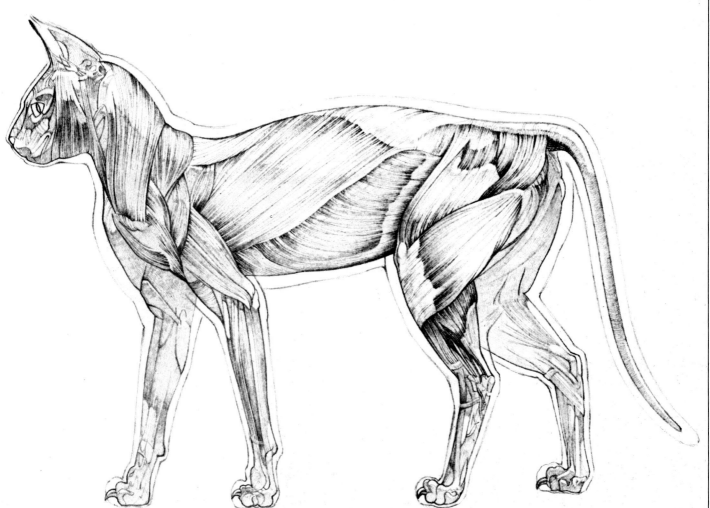

Running and jumping

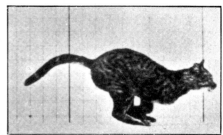

The domestic cat is a lazy animal, and it would much rather find a comfortable spot on your bed, or your favourite chair, than engage in physical effort.

This does not, of course, mean that it cannot do so when it wishes, or when events force it to. As everyone – particularly the unwary mouse or bird – knows, cats are among nature's fastest movers over a short distance.

They are all digitigrade, which means they walk on their toes, rather than the soles of their feet, making them very light of foot. The soft pads, on which the hunting cat can move so stealthily, are also on the toes. The heel bone is well-developed, but is set far back, so that it never touches the ground at all. The pads of the feet form cushions, which act as shock-absorbers, protecting the vital bones on which the cat's weight rests.

Cats cover the ground in a series of giant leaps, and do not run in the same fashion as the dog, or most other animals. They move the front and back legs on one side, and then the front and back legs on the other. The only other mammals to do this are the giraffe and the camel, and these ungainly-looking creatures can hardly be compared with the feline.

The body of the cat has a quite remarkable degree of flexibility. The point of the shoulder is open and free, which allows the animal to turn its foreleg in almost any direction without difficulty. In addition to this, the feline clavicle, or collar bone, is extremely small; in fact some cats do not have one at all.

Cats are finely-tuned physical instruments, beautifully balanced and able to rotate their powerful limbs without the hazard of dislocation. Their overall framework is a masterpiece of elasticity, and this is what makes them so lithe and agile. No part of their bodies is as mobile as the spinal column which allows the tail to bend at will in any direction.

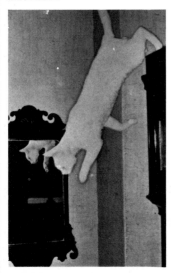

The famous series of pictures of the running cat (top) is from Eadweard Muybridge's Animals in Motion. It clearly shows the grace, power and sheer athleticism that make the cat one of the fastest movers over a short distance. Oddly enough, the only other animals with the same action are the ungainly camel and giraffe. The frame of the cat is a natural masterpiece of elasticity, allowing it to run and jump with ease, using the pads of its feet as shock-absorbers. This protects the bones on which its weight rests.

A balanced view

Cats have a deserved reputation as sure-footed creatures, and yet some will not even attempt a walk along a narrow ledge, fence or the top of a wall.

Kittens may attempt an exploratory step or two, but then reverse to safety. One can, therefore, conclude that such ventures are the result of maturity, and that the confidence to attempt them is tied to experience.

It is rather like children at a swimming pool; they must master the low diving boards, before tackling the higher ones. There have been no controlled experiments to prove the case, but common sense indicates that kittens have to learn the fine art of balance.

Just like humans, cats are subject to hazards and infirmities. They get pregnant, they become arthritic, they grow old and less agile. There are also the tragic cases of disability caused by man's interference. Many additives and pesticides can cause damage to the central nervous system and send the poor creature toppling from the simplest perch.

The normal, healthy cat is capable of performing balancing feats which would leave the average circus performer shaking his head in envy. However, even the athletic feline does occasionally meet situations which make retreat necessary.

How does a cat halfway across a 'tightrope' retrace its steps? If the area on which it is balancing is sufficiently wide, it will turn in the same manner as the performing elephant, and some will look just about as graceful. It balances on its hind legs, then transfers the weight to one front leg, which is placed as far back between the hind ones as possible. The next step is to bring the other front leg around to carry the weight, while it shifts its hind quarters.

There may be many reasons for the cat's decision to make an about-turn in the middle of its journey. It may believe that it has ventured too far from its home base. In the domestic situation, the most common reason for reversal would be an encounter with a fellow feline equally intent on getting across.

This sort of confrontation leaves little time for social niceties, and even less for threats and submission. The rules are simple, and instantly obeyed: the weakest goes overboard, without any argument. Nature's laws do not pretend to be merciful, or even fair. They merely ensure the survival of the species, by eliminating its weakest members.

The cat may slow its descent by hanging on with its claws. One of the signs of this sort of feline trauma is that the claws are scraped down to their bleeding quicks.

If a cat knows that there are obstructions below, it will always try to leap clear. However, an uncalculated fall from any height may upset the animal's reactions. A drop from a considerable height may result in serious injury, no matter how fit or prepared the cat may be.

Occasionally, legs are broken, but more often than not these powerful and resilient limbs absorb the impact with little more than a slight strain. However, the relatively large head of the cat is quite likely to hit the ground, and this can often mean a fractured jaw, or damage to the roof of the mouth.

It is possible to damage the cat's sense of balance through excessive doses of some drugs. The mature cat affected in this way may never get used to the idea that a skill it once could practise with a flourish now completely escapes it. It will go on trying to balance on ledges only to find that it is back where it started on the ground.

Nine lives . . .

'Cats always land on their feet.' This old saying is almost always true, for the animal will make every effort to do so. This picture sequence shows the cat using its tail to right itself while still in mid-air, and make the landing squarely on its feet, so that its powerful legs will absorb the shock.

The cat perched comfortably on the fence (right) is demonstrating the fine art of balance. It seems unlikely that it is an instinctive ability. Common sense must indicate that, just like most other things, the cat must learn how to balance, simply by trial and error.

Climbing equipment

Most boys know that a sheathed knife stays sharp. That is the reason why cats retract their claws. They must keep them razor-sharp to meet every contingency, for these controllable extensions of the body are essential.

In contrast, the claws of a dog are more like finger nails. They may add protection to the digits, but hardly compare with the sensitive functions of a cat's claws.

There is no point in cutting or blunting the claws because you think them too sharp. The cat knows what it needs and will head straight for the nearest tree-trunk and restore the points. It may just use your carpet for the purpose, which will be your own fault. You cannot change a million years of evolution with a few snips of the scissors.

How do cats climb? They use the hind legs to propel themselves neatly and apparently effortlessly as far as possible, and then they scramble. If an irate enemy is in hot pursuit, this initial leap and scramble may carry the cat anything up to about 12 feet.

How do you get a cat out of a tree? The easiest way is to drag it down, but this will entail a few scratches. The best way is simply to do nothing and wait. After a few hours, the panic will subside and the cat will begin the process of inching its way down again.

The retractable claw

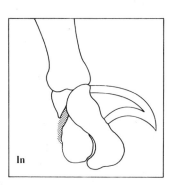

In

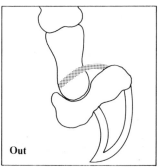

Out

Climbing is part of a cat's way of life, but to do it efficiently, sharp claws are a must. These fully-controllable extensions enable the animal to gain a firm foothold in the most unlikely places. The mechanism of the claw is clearly shown above, both retracted and extended. If fear is the spur and a snarling enemy is fast behind, the panic-stricken cat can use its powerful hind legs to leap high into the safety of a tree, or even scramble up a ladder, an activity which cats usually avoid. It is only when at the top of a tree or some other elevated position, that the cat's natural fear of heights comes flooding back. This will freeze it into immobility, but you can safely ignore the piteous mews if you wish, for the animal will come back down unaided, although this process may take hours.

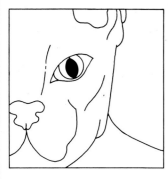

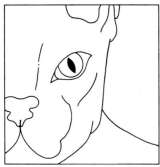

Eyes

Cats cannot see in total darkness. But they do utilize all available light. The iris narrows to exclude much of the light at high noon. The pupils become slits. In darkness, they enlarge to a full circle. The eyes have a layer of cells, the *tapetum*, behind the retina which causes them to shine in the dark. Like other carnivores, cats have eyes which face forwards so that they can judge the distance of prey or enemy.

Ears

The cat's ears can be pricked forward in curiosity, exploration or friendliness, laid backwards in aggression or flat forwards while fighting. They can lie inwards or outwards in submission. They move independently for the cat to focus its hearing. The semi-circular canals of the inner ear control balance. If the cat loses this sense, the vet will suspect ear damage or infection. Cats can hear frequencies two octaves higher than humans can, and sounds of lower intensity.

Nose

The nose is much less prominent in domestic cats than in many other species. Cats are much less dependent on their sense of smell to chase their prey. It is used primarily in feeding, social and sexual activities. The nasal passages are lined with the turbinate bones – delicate scroll-like passages. The cat's nose is easily injured. A chastening tap on the nose may cause a lifetime of trouble.

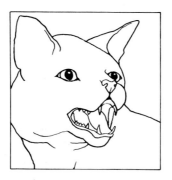

Mouth

A cat may lose limbs, sight and hearing and still survive. But if its mouth is disabled it can neither feed nor fight. The teeth are formed to permit the cat to carry delicate kittens at one moment and mercilessly tear its prey apart at another. From the mouth come those varying sounds which man interprets as seduction, satisfaction, begging, warning, and an expression of pain. The tongue tastes, is used for lapping, and its rough, sandpaper-like surface is used as a comb in grooming and in shows of affection for human companions.

Whiskers

Whiskers are a sensitive means of detection. They approximate to the width of the cat – if they can enter a gap, the rest of the cat can follow. Many a frizzled-whiskered feline will testify that they are most useful in sensing fire. These vibrant bristles emphasize any angry display of aggression.

Paws

Pads are sensitive to pressure. They are pliant, soft and supple. They grip and feel. They are essential for the instantaneous movements characteristic of the cat. Claws anchor the cat to a tree, your carpet . . . and sometimes to you. Paws as a whole are the means of grooming, digging, burying, playing, hunting and exploring.

The five senses

The cat is a highly-sensitive creature with muscles and nerves tuned for an immediate response to pain and pleasure, hunger and thirst, comfort or danger.

The luminescent quality of the eyes has provoked suspicion and, particularly from the Ancient Egyptians, reverence. Their word for cat was *mau* which meant 'to see'. They believed that the cat's eyes mirrored the rays of the sun and protected mankind from darkness. We have all experienced the cat's unflinching stare . . . the Egyptians believed this stare went deep into the human mind. Some say that the cat is colourblind, and only sees shades of grey.

The tail and ears combine to aid the cat's amazing sense of balance. The hearing is assisted by the ears turning like homing devices in the direction of sound.

The fur and skin keep the moisture in the cat's body at the right level. The cat either sheds or grows fur to maintain a body temperature suitable for the climate.

A large area of the cat's brain is reserved to deal with messages from the sensitive pads on the paws. This is one of the few areas through which the cat perspires. When the feline senses a game, it can retract its claws and protect its human and animal friends.

Skin and fur

The skin and fur are efficient insulation and protection against bites, scratches and blows. They are waterproofing, camouflage in some situations and a threatening apparition in others. They are sensory organs that can register pain, pleasure, heat and cold in instantaneous reflexes. All that plus sheer beauty. Have you any suggestions how nature can improve on the covering she gave our cats?

Tail

The tail forms part of the cat's spine and is necessary for balance. It is an expressive organ and can say 'hello' or threaten destruction. The tail is a meteorological instrument and can tell the temperature and the direction of the wind. Tail movements accompany any high level of excitement – whether in sexual or hostile encounters. In the 'social' dog the waving tail is usually a friendly sign. In the generally 'non-social' cat it is probably an antagonistic signal. And we all know what it means in the rattle-snake. The Manx cat has an hereditary defect in that most of its tail does not develop, but it doesn't seem to miss it.

The life-cycle of your cat

Kittenhood, the full years of maturity . . . and leisurely old age.

The first days

The newly-born kitten and the fading great-great-grandfather have two things in common. They are both relatively helpless. The kitten has the advantage of a doting mother to guide it into early adolescence. The very elderly cat has the slight advantage of experience. And they both have relatively large heads which tend to wobble. This is common to all new-born kittens and aged cats who die of nothing other than old age. How can that short and skinny neck which contains seven cervical bones carry the weight? The first two bones are so modified that the cat can turn its head from side to side as easily as it can turn it up and down. And they are supported by muscles that any other species would envy. It seems miraculous that a mother cat can carry a kitten that weighs almost as much as she does over big fences and for considerable distances, but physiologically they are well-geared for that sort of task.

The newly-emerged kitten has limbs that are about as useful as fins. It flounders and paddles, a reminder that we all evolved from creatures that once lived in water. The kitten gradually becomes more sure-footed. Its limbs begin to act like arms and legs, but they are still relatively pliant. The logic of that evolutionary plasticity is simple. The growing animal is prone to all sorts of accidents. Pliant bones don't fracture easily. And, if they do, they heal very easily. The exceptions are those cats which are not allowed a natural diet which includes bones or bone meal.

During the middle stages of development – late kittenhood, adolescence and maturity – the cat's shape gradually changes according to geometric principles. The athletic youngster is roughly triangular, with a broad back which tapers into a very tiny abdomen. All muscle up above and a tiny abdomen underneath. Later, the adult becomes rectangular. During late maturity and merging into old age, the muscles at the top tend to waste and the waist below becomes more prominent. The triangle has become inverted. This simple rule of thumb is an aid to telling the difference in ages between cats, and shows a cat that has prematurely aged through serious illness.

The face of youth . . .

1

. . . and of age.

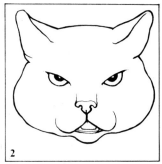

2

The shape of the kitten's head (left 1) changes dramatically as it matures (2). The ears become more pointed and move upwards from the side of the head. The face become more stretched and, in the case of the tom cat, widens around the jowls.

The five ages of cat (right) make an interesting study. Just as in Shakespeare's slightly longer version dealing with man, the cat at first and last is a relatively helpless creature. The newly-emerged kitten's limbs are about as useful as fins (3) and it can only drag itself about. As it grows up (4) it is much more sure footed. Its body shape changes as it reaches adolescence (5). This athletic youngster has a triangular shape, all muscle above and a tiny abdomen below. Later, the adult (6) becomes rectangular. Old age (7) brings a complete reversal as the triangle of youth is inverted, the muscles at the top tending to diminish and the waist becoming more prominent.

The very young kitten

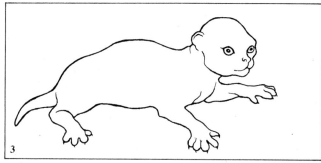

3

The kitten

4

The adolescent

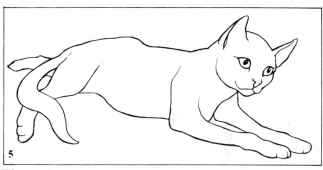

5

The mature cat

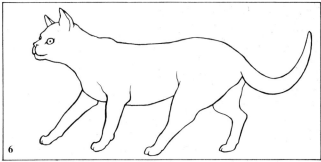

6

The very old cat

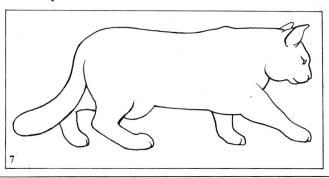

7

The suckling kitten

The kitten at birth is deaf and blind, and almost completely helpless. It can move only a limited distance by using a sort of rock-and-roll shuffle accompanied by a characteristic head shaking.

However, it has an acute sense of feeling. The mother licks the kitten and by instinct it orientates itself from the mother's mouth towards the nursing area. The kitten crawling about in a haphazard manner usually has first contact with the outstretched legs of the mother while she is lying supine and follows the legs upwards until it finds a nipple.

The feeding behaviour of the new-born kitten has two instinctive elements. The first is the search for the nursing area and the second is the response to the nipple. During the first few days the response seems to be to the same teat. Later on the growing and every more hungry kitten becomes less fussy.

There are three major phases in the relationship of the nursing kitten and its mother. During the first phase the mother initiates the whole feeding process. During the second, when the kittens are more independent, the relationship is more reciprocal. The kittens waddle towards the mother, and she assists the odd laggard. During the third phase, the kittens actively demand to be fed. They will follow the mother about until she stands or lies and allows them to feed.

Even at this last stage, the most callous of mothers will quickly respond to the pitiful call of a kitten which has wandered. She will not rest until she has found it and carried it to a place of safety.

Not until their eyes are open do the kittens have any semblance of independence, and even then they will need their mother for some weeks to come. Fortunately for them, her maternal instincts are very strong and they reciprocate with an equally strong instinct to survive.

How mother helps out

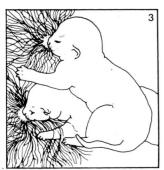

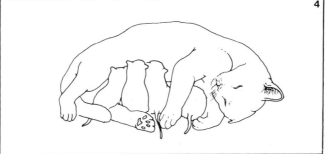

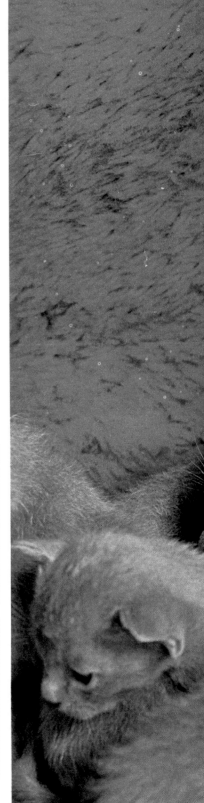

1 A nursing mother may reject one or more of her litter which appears to her to be sub-standard. This rejection is almost always irreversible. Nothing will persuade the mother to accept a once-rejected kitten, and the hapless owner must undertake the round-the-clock job of hand-rearing. However, she may accept and rear a creature quite unlike herself, like a puppy. This behaviour, stimulated by hormones, is a physiological not a psychological reaction.

2 Many animals, including kittens, lick their mother's mouths. In this way, the mother passes on antibodies which help to fight disease. The habit also reinforces the links between the mother and her young.

3 and 4 Sometimes kittens seem to burrow underneath the mother. They're not going to suffocate . . . they know what they are doing. The mother has two rows of nipples, and each kitten knows on which side it customarily feeds.

Learning behaviour

Learning behaviour

No kitten can be expected to go from the coddled security of milk and nest directly into the skilled activity of hunting for food.

During kittenhood and adolescence, the young cat learns the skills that in maturity become second nature. The highest form of learning is *insight learning* which is the conscious consideration of a problem. In the laboratory it is difficult to devise experiments which simulate problems in the field, but it can be shown that higher animals, especially monkeys and apes, are capable of insight learning. Chimpanzees, for example, will pile boxes one on top of another so that they can reach a reward of fruit. Cats do not learn through insight; their instinctive behaviour is modified through experience, habituation and inhibited play.

It is difficult to distinguish between learned and instinctive behaviour. A litter of kittens may be left on their own for some hours while the mother goes off hunting. If they are left in a relatively cold nest they will instinctively cuddle together. The group can keep itself warm whereas a kitten on its own might rapidly lose body warmth and die. The same litter, left in a very warm area, will spread themselves out. This is almost certainly instinct.

A kitten learns through experience, or *trial and error*. This means that if a cat accidentally performs an action which is rewarded it may then deliberately do it again in the expectation that the reward too will be repeated. This is how many cats gradually learn the difference between the many possible expressions of their family of fellow cats, dogs and people. They will learn that every time they approach an individual who is not interested in play they receive a painful bite, scratch or kick. After a while, even the most energetic kitten learns that it is best to pause and take stock before rushing in. Similarly, a playful kitten, while jumping about, might accidentally open a cupboard door. It may learn to open the same cupboard door by imitating the mother. A whole world of forbidden tidbits is revealed. By trial and error the cat learns to avoid punishment or receive reward.

Habituation is the simplest and most common form of learning. This results from the loss of an old response. In the same way that crows tend not to be frightened of scarecrows, the cat soon learns that falling leaves or shadows carry neither reward nor punishment. They are things which can safely be ignored.

It is quite obvious, when a kitten plays with a ping-pong ball, that something important is being learned. The kitten makes a tentative approach and gives the strange object a small pat. The ball moves and in effect is saying to the kitten, 'I am running away and therefore I must be chased.' Usually, within seconds, the kitten learns to dribble the ball. Scientists call this *inhibited play*. Later, the kitten will go through exactly the same motions while pursuing a mouse. At this stage, it knows it should keep the object in control, but has neither the knowledge nor the experience for the quick kill.

In the cat's feral state almost every outsider is its potential enemy. This certainly includes men and their canine friends. So why do kittens reared in a highly-urbanized society ignore these primitive 'enemies'? They have learned through their mother's lack of response that these creatures represent no threat. They may be welcomed or ignored. In this way the animal's instinctive behaviour has been tempered by what it has learned.

Weaning

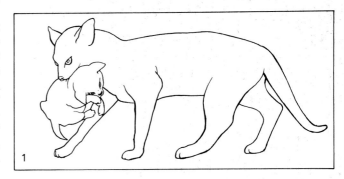

Grooming

1/2/3 The human-orientated cat may have chosen to rear her litter in the sleeping quarters of the apartment or house in which she lives, despite the fact that the people who open the tins and grill the steaks continue to feed her in the kitchen. No matter how clever the cat, she is simply not designed to carry a bowl of milk. The cat pictured above has solved the problem. She carries her litter to the bowl and they follow her example and lap the milk provided for them.

4/5 When the kittens are very young the mother can groom them at will. Just the action of her tongue is sufficient to control their playfulness. As the kittens grow she may have to use her powerful and flexible front paw to restrain them. The cat uses these paws not only for the gentle art of grooming, but for more strenuous tasks like holding down mice or rats while considering the next move.

Beginning to explore

The kitten must learn the limits of its own body by its own efforts. Where is that sound coming from? Is that strange object worthy of note? Can I move fast enough and pounce quickly enough to catch and not do myself an injury?

The spirit of adventure

Ogden Nash once said, 'The trouble with a kitten is that it grows up to be a cat.' The period in between is called adolescence.

As with young humans, this difficult time may be protracted by a mother who is over-protective, or cut short by her flinging the youngster into the adult world.

The kitten in a good home can take as long as six months to learn the game of life. The less fortunate grow up much more quickly. Skills learned in play must soon become established routine for simple survival. Like underprivileged children, neglected kittens have neither the time nor the place in which to enjoy luxuries.

In bodily terms, the adolescent cat is certainly one of nature's great successes. It is aware, alert and totally receptive. Its reflexes will never be better and its sensory apparatus works in perfect harmony. All that is lacking is adult strength of muscle and, of course, experience.

At this stage, it will start to make tentative forays into the neighbourhood. The adolescent is on the edge of a vast, unknown world and its steps will be exploratory and highly inquisitive, but it does not have the confidence to wander too far. This wary animal always seems to keep one eye on the rear view mirror and is very easily frightened away.

In the domestic situation, the adolescent is unlikely to challenge established neighbourhood cats for either territorial or social rights. At this stage, it is still obliged to rely on the charitable tolerance granted by adults to the offspring of their own and other species.

The transition from adolescence to adulthood is often marked when the kitten does stake out its own territorial claim. Its elders and betters are then left in no doubt that 'this is my place . . . get out and keep out'.

The growing cat has all the energy requirements of the active adult, in addition to the needs of the immature body. Many owners know only too well how expensive this can be. There are well-authenticated reports of healthy adolescent kittens who swallow half a pint of milk and two eggs for breakfast, a large tin of dog food for lunch, steak for dinner – and then sit begging at the family dinner table.

Male adolescents develop an early interest in the opposite sex, although they may not be able to do anything about it until after their first birthday. They will indulge in the most outrageous, and fruitless, display activities to attract the attention of the queen. Nature does not similarly restrain the young female, for she can quite easily become pregnant at the age of only four or five months.

Life is an adventure for the adolescent cat, but it is still a game, and will be treated as such at every opportunity.
Below: three examples of the growing cat's attitudes to the art of grooming. It is quite capable of grooming itself (1), although it has not yet gained the poise of the adult. At this stage it will also groom another youngster (2), in the hope that it will respond (3) by joining in a mutual wash.

Learning to groom

The thinking cat

The mature cat displays all those qualities of independence and self-sufficiency that its human family finds so compelling. It spends less time playing and more time in activities that have a clearly defined purpose. With adulthood, the cat gains confidence and tends to fall into an ordered routine of day-to-day living. It accommodates itself to its household's way of life and the limitations of territory.

Cats gradually get to know their neighbours and they evolve a social order. One male cat will wait at a crossroads for another to pass. This could be a recognition of territorial rights; it could also be a sign of respect for superior strength.

Cats recognize and tolerate those that they know. Strangers are fought off. They soon learn that the fat, old King Charles spaniel next door can be safely and completely ignored. The nasty little terrier across the street, however, must be kept in a state of constant fear. Each and every time it shows its ugly, snarling face it must be met by a formidable display of ruffled fur, enlarged tail and a threatening hiss that can be heard in hell. Further down the street lives a tired old tortoiseshell which likes to bask in the sun while lying on the window sill – she might respond to a greeting in passing by wearily lifting one eyelid.

The cat has acquired skills and techniques in order to adapt to its environment. As well as protecting itself and its territory the cat now hunts. The first step, of course, is to gain access to the hunting grounds. This poses no problem for farmyard cats or their feral cousins. But the hearthside cat may have to wait patiently by the door to the basement or garden for several hours until it is given the opportunity to make a furtive dash to begin an expedition.

Cats learn in kittenhood that certain actions result in either reward or punishment. Some are particularly quick to learn that by swinging on door handles, for instance, doors open, making hunting grounds more easily accessible.

The mature cat may take control of its own existence. Many are the tales of cats being fed in more than one household, exacting affection and attention wherever they go. The family with which a cat normally lives will be greatly concerned when the pet stays away from home for more than twenty-four hours. It may have been in some difficulties, but more often it is exercising that independence which is so attractive. A feline assignation or an equally warm fire elsewhere, a moonlight hunting trip or extra-rich morsels from a neighbour's table are just some of the possible excuses which it might give.

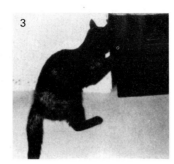

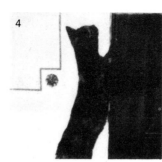

The hunting instinct

1 One obvious benefit to the mature cat of early exploratory behaviour is physical fitness. The muscles become stronger. The lungs increase in capacity and efficiency to accommodate the needs of the muscles. The heart and circulation respond to the increased demands. And, of course, reflexes become razor-sharp.

2 3 4 The mature cat shows evidence of thinking processes which enable it to achieve its demands, and 'cupboard love' is just one example. Many cats learn by trial and error how to open a door. None, however, has the good manners to shut it as it cheekily waves goodbye with its tail.

5 6 Hunting is a major activity. Much of the cat's play as a kitten and adolescent has lead to finely-honed instincts for the kill. It may not need its prey in order to feed but the veneer of domestication is sufficiently thin for it to continue to exercise these drives and ensure the continuation of the species.

Old age

The average lifespan of the domestic cat is considered to be about 11 or 12 years. However, this average is increasing as infectious diseases become more readily controlled and treated.

Many cats are old at the age of seven. At the other extreme, there are cats in their twenties which appear to be as lively as they did in their younger days. Any article in the Press about long-living cats will provoke a spate of letters from people claiming to have cats of 28 and 30.

A high proportion of such letter-writers always seem to be elderly folk who feed several visiting cats, indeed colonies of cats. Without wishing to seem unkind, the likelihood of any cat surviving for 30 years is remote.

Man spends an average of one-third of his life growing up, a third as an adult, and the remaining third in the gentle decline into old age. Cats, on the other hand, spend only about a tenth of their lives growing up, with eight-tenths in vigorous maturity, and a further tenth in the final slide.

Animals born to long-living parents are more likely to live longer themselves. It is also thought that a kitten born to a relatively old mother may have more defects and a shorter lifespan than those of younger mothers.

It is obvious that a cat with a stable home, regular meals and inoculations against disease, is more likely to reach a ripe old age than the unfortunate alley cat.

The vet can make a quick assessment of the elderly cat's condition by simply glancing at it. There are four signals that will be obvious before the patient has even been settled on the table: general alertness, muscular tone, brightness of eye, and condition of coat.

Closer examination will reveal the other signs of old-age deterioration, such as a dry, scaly nose, receding gums, blunted claws, harsh coat, and difficulty in manipulating the spine, head or neck.

This cat will move slowly and reluctantly. Instead of leaping onto a chair, it will either climb slowly, or wait for a trusted human friend to lift it.

Many owners simply do not notice any of these signs, mainly because the disabilities of old age do not seem to affect the cat's appetite. In fact, many senile cats eat with more gusto than ever.

Elderly cats tend not to groom as regularly as they once did, because stiffening joints make the process difficult, and because self-interest seems to lessen with the advance of the years. The owner must give the cat regular grooming, to help it restore its own pride in appearance. Also, the diet must be high in protein to maintain the wasting cells.

As their joints stiffen, so do their minds. Change of any sort is resented. Therefore, it is useless to try to brighten up your elderly pet's life by introducing a kitten or puppy. The cat will hate this.

Moving house is, of course, a real trauma for the aged cat. As long as familiar objects are available, it will settle comfortably within the confines of the house. It is the outdoor aspect of the move that causes problems. In its old home, it probably spent several years establishing superiority over the neighbourhood animals. It cannot do that in its new environment, for it is sure to lose any territorial fights with younger cats.

The two criteria for euthanasia are: First, is the condition making life a misery? Second, can the condition be cured? Fortunately, the vet has a whole array of drugs which can ensure that the solution to a negative answer is painless.

The life-style of your cat

Its activities, rituals and habits.

Cap-napping

Whereas human beings spend about a third of their lives asleep, the cat averages more than sixty per cent. Scientists do not know the causes or the functions of sleep but they do know that prolonged interruptions of the cycle of rest periods can produce illness. They also differentiate between five different kinds, ranging from a light, sensory drift or 'floating', through light sleep to three degrees of deep sleep. The first is the phenomenon of catnapping, which is so well recognized that the term describes that form of human sleep.

During light sleep the cat moves its limbs frequently. In deep sleep it will move very little, except for the eye muscles and the ends of the limbs, but the electroencephalogram will look very similar to that of the cat when it is awake. This form is known as *paradoxical sleep*, and this is when dreaming occurs. The adult cat spends about 15 per cent of its life in paradoxical sleep. Kittens, immediately after birth, spend over 90 per cent of their sleeping time in this state, but by the time they are four weeks of age they are down to the adult level. If cats are robbed of their paradoxical sleep for three or four days, they have to compensate for it by sleeping longer and deeper at the first opportunity.

Many sedative drugs can alter the character of paradoxical sleep and cats appear to be more sensitive to these drugs than other domesticated animals. They should be used only when absolutely necessary. Perhaps the sleep they induce changes dreams into nightmares.

A little shut-eye

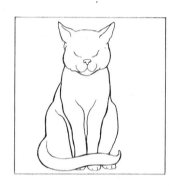

Many cats choose old trophies or half-open drawers to sleep in, along with the traditional spot on the family armchair (far left). This is because, unlike dogs which flop down anywhere, cats need security before sleeping. Some exotic fish secrete a cocoon around themselves before they go to sleep; the cat will go to considerable lengths in order to emulate this, using objects in the domestic environment to achieve it. This does not apply to catnapping (above), when any perch will do.

Waking up

Home-bound cats often follow the routine of the people with whom they live. Despite the seasons and the weather, alarm clocks ring and activities begin at the same hour each day. The cat soon learns that all this jangling activity is the prelude to its breakfast. That is, for five days of the week. But there may be two days of the week that this frantic bustle does not happen. No matter how hungry, or how piteously it miaows for its breakfast, the whole ritual is postponed for no clearly decipherable reason. The mature cat knows that it might just as well roll over and go back to sleep.

A common definition of sleep is that it is that state in which activity and interaction with the outside world decrease or indeed stop altogether. The obvious bodily changes that take place are that both the heart and respiratory rate become considerably slower. Some animals sleep so deeply during hibernation that it is not always possible to tell, even with the aid of a stethoscope, that the heart is beating. Many a child has brought a 'dead' tortoise, hamster or hedgehog to the vet for disposal only to be assured that it is only sleeping very deeply. At the other end of the scale, there are animals like antelopes that hardly sleep at all. They have to be ready to escape at all times. Some animals remain vigilant while others in their community rest. The swift manages the clever trick of sleeping while flying – during gliding it can catch the occasional nap.

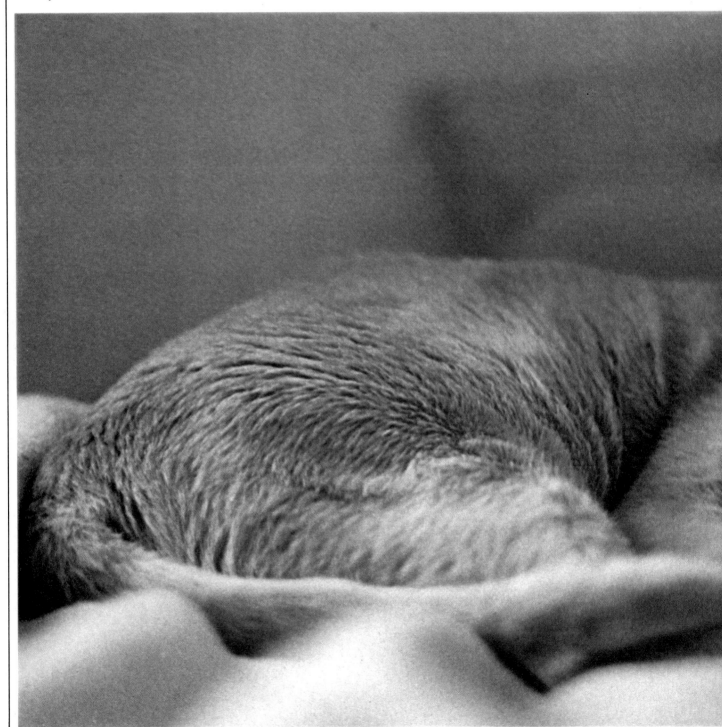

The temperature of the body drops slightly in warm-blooded animals. Bats become almost cold-blooded during sleep. When they wake they must stretch and exercise to raise their temperature to a normal level. Similarly, the waking cat (if it has not been startled) will go through a leisurely ritual of yawning and stretching. These yoga-like exercises are performed with a thoroughness that relatively stiff-muscled humans envy and try to emulate. Every single joint from the top of the head to the tip of the tail appears to be moved. This has the function of restoring full circulation and instant readiness for action to every part of the body. It may also be one of the ways in which an apparently sedentary creature keeps itself superbly fit.

Ritual performance

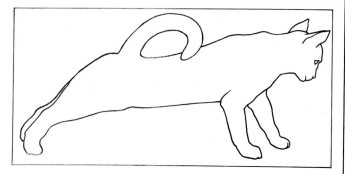

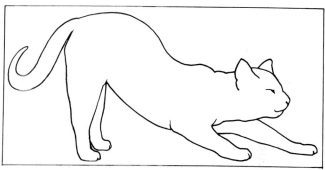

Most humans make do with one all-embracing stretch. Cats usually embark on a whole series of movements, designed to loosen almost every area in the most pleasant and satisfying manner. This process is presented as a public performance, unlike so many feline activities, and it will usually commence with a joint and muscle-loosening extension of the front legs. No respecters of such trifles as your upholstery or carpets, cats will dig their claws in to give themselves an effective anchorage. This is followed by the arching of the back, in which the animal squeezes itself into an amazing and concertina-like posture, before completing the acrobatic and, at times, almost balletic spectacle. The finale is usually reserved for the rear legs, which are each stretched out in turn. The whole show can be accompanied by a further selection of face-twisting yawns. When this ritual is over the self-respecting cat feels able to face the world.

The feline gourmet

Early man used to throw left-over food and bones at animals lurking outside his cave. His intention was to discourage them, but it merely convinced certain animals that man could simplify the eternal quest for food, by handing it to them, almost on a plate.

The cat was among those animals and since then it has moved right into the home. However, its approach remains unchanged. When meat is being prepared, children are eating cereal, or a cow is being milked . . . the cat is there. Cat-lovers will recognize the obvious ploys, such as piteous mewing or dog-like begging.

Even in rich, urbanized societies, where most pets are overfed, cats persist in begging. One explanation is that the cat becomes addicted to eating, just as some people are hooked on nicotine or alcohol.

How much should a cat eat? A simple rule is one ounce of food per ounce of body weight while growing, and half that amount for the adult. Like humans, cats vary considerably in their ability to utilize food, and in their daily expenditure of energy.

If your cat is overweight, take it to the vet to make sure there is no underlying physical cause. If the obesity is because of overfeeding, cut the food intake by five or 10 per cent, until it is down by 50 per cent.

Because cats are covered with hair, it is sometimes difficult to judge weight. Also, owners tend not to notice gradual changes. If the adult cat gains only four ounces a year, by the time it reaches an age of 10 or 11 it may be a quarter to a third overweight.

The cat's traditional diet is the small rodent. The mouse is a real meal, providing 70 per cent water, 14 per cent protein, 10 per cent fat, one per cent carbohydrate and one per cent mineral. The liver is full of vitamins and the bones contain calcium. The ideal manufactured cat food should approximate to that formula.

Apparently normal cats, eating the recommended diet, will still dash straight from the bowl into the garden and eat the grass. This behaviour is not confined to the domestic cat. Lions in game reserves can be seen in the mornings grazing like cattle. Although no one knows for sure why they do it, one can only assume that fresh grass contains nutrients lacking in the regular diet.

Cats also eat grass in an attempt to soothe inflammation of the throat, or when they have an irritation in the bowel. This type of grass-eating is characterized by a compulsive, almost frantic, urge, the cat seeking out coarse, dry shoots. If the unfortunate animal has swallowed a noxious substance, or has accumulated some hairballs, this is an effective way of inducing vomiting and helping to relieve the discomfort or distress.

The rib test

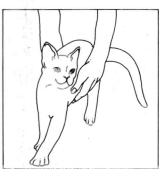

The rib test remains a reasonably effective guide to whether or not a cat's weight is correct. If you cannot feel each rib individually, without exerting undue pressure, then the cat is probably too fat. If the ridges of the cat's ribs feel like pencils, the chances are that it is not being fed enough, or that it is suffering from a condition which needs immediate attention from your vet.

The scavenging cat

Stealing is basic to a cat's way of life, and this phenomenon runs right through the feline social scale, from one end to the other. The outcast alley cat has an obvious need to scavenge in such likely places as dustbins, in order to supplement the otherwise meagre diet of anything it can catch.

However, this behaviour is mirrored by the fattest domesticated cats. This pampered creature is certainly not above wobbling away from a bowlful of food and a dish of cream – and then stealing the family's dinner while human backs are turned.

Cats will go to considerable trouble to take food, even though it is not needed. Some experts say that it is sheer force of habit. One slightly-puritanical theory suggests that some innate part of the cat's make-up dictates that no reward is possible or acceptable without some effort.

Quite simply, it is said, the cat cannot really enjoy any meal that it has not begged, stolen or hunted.

Whatever the truth of this, it is undeniable that cats can be very fussy eaters. Unless introduced to a varied diet at an early age, they tend to have ultra-conservative tastes.

That is not to say that the adult cat is inflexible. Most animals, placed in a cattery while the family is on holiday, refuse all food for around 48 hours. On the third day even the fussiest cat will usually eat whatever is offered.

The response to this situation may be sheer hunger, but it is the sight and sound of other inmate cats gobbling the food that arouses the primitive competitive spirit. 'If I don't eat this stuff, disgusting though it is, one of those strangers might get it.'

Although cats are among nature's greatest carnivores, it is possible to convert them to vegetarianism, if the vet advises such a radical dietary change. Simply add portions of the new diet to the old and, over a period of around three months, phase out the undesirable food content.

The wages of sin?

It has been said by certain theorists that to a cat no meal is really worth the eating unless it has involved the eater in at least some token effort to beg, steal or hunt. Certainly, even the fattest of domestic cats will go to considerable trouble to steal, even if it is merely knocking down and ripping open a box of its own dried catfood to win the unambitious reward of a few illicit morsels.

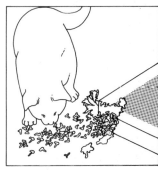

Tastes in drink

Cats are private creatures, and many owners never even see their pets drinking. Every day they put down a bowl of clean, fresh water – only to remove it, still untouched, 24 hours later.

This is not to suggest that cats have discovered phenomenal survival powers. They can tolerate weeks of starvation, but but only a few days of dehydration will kill them.

The average cat needs about 40 teaspoonsful of fluid every day. Growing cats require as much as a daily 10 per cent of their body weight in fluid, while the adult can make do with only half that amount.

The most common source of fluid is the food a cat catches or is given by its human family. In addition to this, most enjoy, and are regularly given, milk. The exceptions to this are adult orientals, such as the haughty Siamese, which do not care for it.

Although the phenomenon is rarely observed by human eye, some cats make a habit of dowsing and kneading their food in water before eating it. There is no generally-accepted explanation for this, but the action obviously increases the animal's fluid intake. The raccoon, for example, gets much of its food from streams, and dowsing is a normal part of its behaviour. The link here is that cats and raccoons may once have been related species.

As a rule, cats do not like water which has been treated in any way. Chlorine, flouride or just old lead pipes, no one can pinpoint the reason for this feline distaste. Cats are perverse and, if offered that kind of water, they will sniff and walk away, tail disdainfully high in the air.

However, exactly the same water may be entirely acceptable if it is provided in running form. The reason for this may be that such water is more fun to play with than to drink. Whether it is just an irritating drip, or a powerful gush from the kitchen tap, the cat will spend hours partly playing and partly drinking.

More than one cat has learned the art of turning on a tap or pressing the appropriate lever. It does, of course require a good deal less intelligence to open a bottle of milk – although the cat living in the United States would probably need to open a carton.

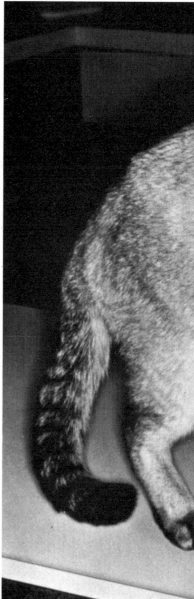

Cats are perverse creatures and may refuse water put down for them in a bowl – deciding instead to lap up the drips from the kitchen tap. Many of them have even learned to turn on the tap. Similarly, the business of removing a milk bottle top is nothing to the average thirsty cat.

Self-grooming

The first active relationship between mother and newborn kitten is grooming. Within seconds of birth, the mother is using her tongue and teeth to loosen the kitten from its enveloping membrane, stimulate breathing and dry its soaked body.

Within hours, the mother must remove parasites such as fleas. These congregate in dozens around the mouth and eyes of the helpless kitten and, unless the mother is diligent, the kitten's strength will be sapped. Many kittens are permanently stunted by fleas which the unfortunate mother could not control.

Grooming is one of the first activities that the kitten is capable of performing on its own. In one study of 40 kittens under laboratory conditions, the eyes opened at 12 days, walking began at 22 days and grooming a day later.

Kittens reared without their mothers, in separate incubators, began grooming at 15 days, two days before they began walking. It was difficult to evaluate, for the kittens tended to perform sucking movements on their own bodies soon after birth.

Grooming actions are inborn. Some species of rat wash their faces using the palms of both paws. The paws make effective brushes because the palms are flat. The cat cannot use its paws like this, because its claws are too long and the pads too closely set.

It uses its forearms as a brush, and then uses its tongue to clean the forearm. Unlike many smaller animals, the cat uses only one forearm at a time while grooming. Obvious anatomical limitations confine that sort of grooming to the face and head. Most of the grooming of the other parts of the body is done using the tongue alone. The fit cat can assume the most unlikely and seemingly unbalanced postures with an enviably athletic grace and style.

Throughout the entire animal kingdom, the process of grooming occupies only slightly less time than sleeping, or even mating.

Furred animals have little option in the matter, for they must constantly strive to keep their covering in the very best of condition. This is partly for comfort, but its more important function is to maintain its unique qualities of insulation.

One reason for grooming is the sheer pleasure it gives the animal involved. Most cats come to enjoy being groomed by their owners. The areas they particularly enjoy having scratched or stroked are those that they cannot easily reach or deal with themselves.

The meticulous wash

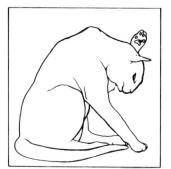

Grooming is an instinctive activity for cats and kittens, and some begin before they can even walk. They use their forearms (left) to wash their faces, licking the forearm to clean it after brushing the face. Above: All areas inaccessible to the paws must be groomed with the tongue and teeth. Cats can reach a surprisingly wide variety of contorted positions with grace, agility and, most important, dignity. Right: Sheer pleasure is an important factor in grooming and, if they cannot be scratched, rubbed or stroked by their owners, cats enjoy grooming each other. In common with other animals, particularly monkeys, one adult cat will approach another and 'solicit' to be groomed. This behaviour is known as *allogrooming*. The human groom will be rewarded with a contented purr . . . and the occasional itch.

... and a helping hand

The easiest way to groom a cat is to get one that does not need any grooming. This is not as facetious as it sounds, for the majority of cats can and do maintain themselves without any human assistance. It would be arrogant to presume that creatures which have managed by themselves for millions of years have, in a mere 5,000 years of domestication, become dependent on man for this essential activity.

The great exceptions are thick-coated and long-haired types which man has encouraged with selective breeding. Perhaps the most beautiful of these are the Angoras, or Persians. Most breeders and lovers of these types proudly admit that it requires at least 30 minutes of daily grooming to keep them in prime condition.

Every vet, boarding cattery and feline beauty parlour will agree that only a few months of neglect can create such a tangled mess that the only method of restoring order is by cutting or clipping off the whole lot. Nor is that an easy task, for the animal often requires a general anaesthetic.

To prevent that situation arising, the owner must get the kitten accustomed to daily grooming from the day of its arrival in the home. As with children, the introduction must be painless and, if possible, pleasurable. Many a cat which has been gently introduced to combing and brushing will eagerly leap onto the table and purr throughout the proceedings, no matter which part of the body is being combed.

There are hundreds of grooming aids on the market, many of them indistinguishable from those used on dogs. The majority of the aids are combs, brushes or gloves. Finely-spaced combs – known in the trade as flea-combs – may be useful in grooming short-coated or silky-coated cats. However, their use is limited on cats with more luxurious coverings, for they tend to glide over the surface without ever penetrating to the layer below. Although they are called flea-combs, it is important to remember that they do not automatically eradicate the target. Unless the flea is trapped and killed, the comb's only function may be to give the parasite a free ride to another part of the body.

Before undertaking this painstaking task, the owner should remember that no matter how numerous the enemy, the only really effective way of killing is to focus on one flea at a time. Follow three, and they all go free.

One of the most useful combs is the one shaped like a garden rake, but with two rows of parallel teeth fairly widely spaced. This tool is so under-rated on the market that many good pet departments and shops do not even bother to stock it. However, it is well worth the effort involved in looking around for one. This invaluable aid can easily and efficiently penetrate and remove dead hairs from the thickest-coated animal.

It is common knowledge that cats do not enjoy taking a bath. Indeed, most of them do not need to, for they will constantly wash themselves. However, even the sure-footed cat can occasionally fall into something that has to be bathed away. In this event, place the cat in two or three inches of water, soap, rinse and repeat. Be sure to keep first aid equipment handy to treat your scratches and other wounds, for your cat can be relied on not to feel the slightest gratitude for your assistance.

Curious cats have been known to fall into the bath. If they get soaked to this extent, they have great difficulty in washing themselves dry. A very gentle going-over with the hair-dryer may frighten a wet cat for a moment, but the creature will soon be basking in new found comfort.

A guide to grooming

Untangle compressed areas of hair with a comb, and then brush the coat against the grain. Complete with a final soothing brushing or massage with a coarse glove. However you do it, regularity and frequency are essential. Some long-haired types require at least 30 minutes of grooming every day, in order to avoid a tangled mess that often requires cutting off altogether ... and that may mean a general anaesthetic in a surgery.

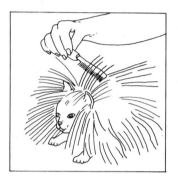

Combing should be a regular feature of the cat's daily life, from the time of its arrival in your home. Never use a sharp comb or brush on the tender areas of head and stomach. Always make sure that you are using the right comb to penetrate the lower layer and remove any dead, matted hair. When using a flea-comb, remember to single out only one flea at a time. If you try for a massacre, they will all escape and live to fight again.

Difficult clumps of hair may be scissored away, but do not attempt to do this job unless you have a pair of curved scissors. Even with that safeguard, do not snip until your forefinger and thumb are between the blades and the cat. Once you have cut off most of the knot, the rest should be teased apart with ease. If possible, have an assistant to hold and soothe the cat, or you may have to deal with an angry protest, and need first aid.

A cat's introduction to grooming must be painless, soothing and pleasant. If that is achieved, even the short-haired cat will enjoy regular grooming sessions, and may even eagerly leap onto the bench or table. The cat will then purr throughout the procedure and co-operate no matter what outlandish position it is being asked to adopt, or which part of its body is about to be groomed. Brushing and smoothing can be a feline fetish.

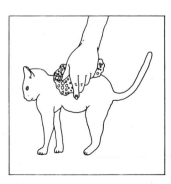

Occasionally, even the best-tended cat has severely knotted hair. This can be soaked with a sponge to help tease apart the knots. Most cats do not need a proper bath, because they constantly wash themselves. If it is absolutely necessary, place the cat on a towel in three inches of lukewarm water, soap, rinse and then repeat. Following this, you should bathe your wounds and apply antiseptic to your many scratches and cuts.

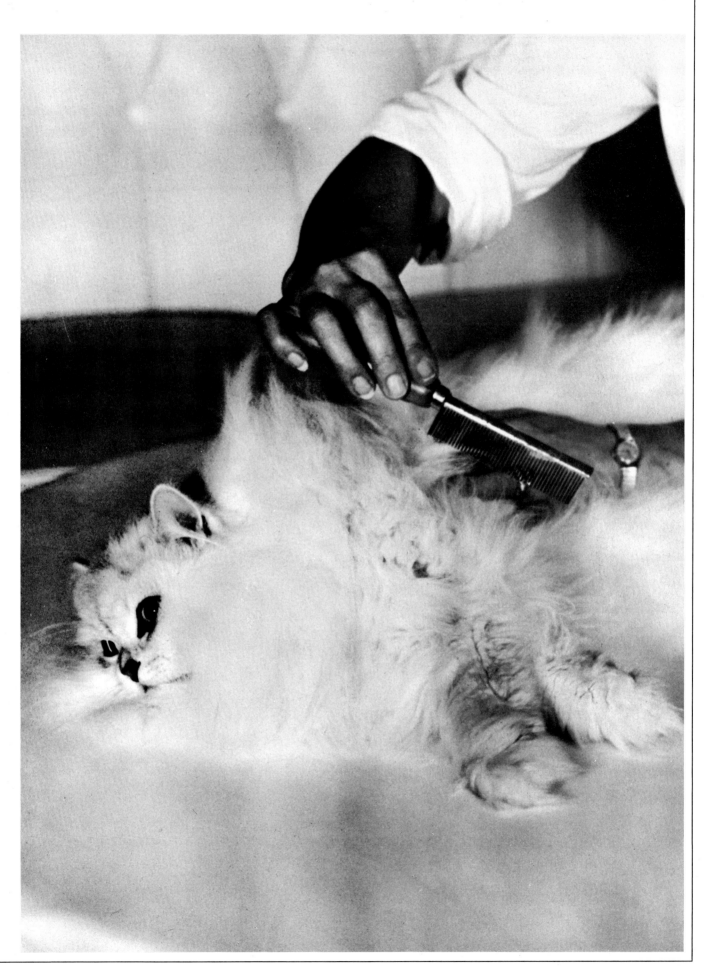

Physical fitness

If your cat displays any signs of illness, or any untoward behaviour which might suggest sickness, take it to the vet straight away. No one wishes to bother the vet needlessly, but it is better to be safe than sorry. Some possible signs of trouble are indicated on this page.

Cats are not usually hypochondriacs, and a caring owner should soon notice if there is anything wrong. With a good home, and the exercise provided by some outdoor life, your cat has a good chance of remaining as bright-eyed as the character on the right.

Some possible signs of trouble

Difficulty in walking. If your cat appears to have a broken limb, its movement should be restrained. The easiest way of doing this is simply to put the cat into its basket – then take it to the vet. If there is no obvious fracture, check that the cat hasn't got something caught in its paw or claws. If there is tenderness, bathe it in warm salt water.

Toilet difficulties. If your cat appears to be constipated, or has difficulty in passing urine, take it to the vet.

Balding patches. These may indicate ringworm. Wash your own hands. Avoid handling the cat, and take it to the vet. Cats are also subject to fleas, lice and ticks. You can cope with these yourself, with initial instruction from a vet if necessary.

Worms. If you have any suspicions on this count, talk to the vet.

Digestive problems. If your cat persistently vomits or has diarrhoea, stop feeding it, and take it to the vet.

Ear Problems. Many cats scratch their ears, twitch them, and shake their heads occasionally, but constant repetition of these actions suggests the presence of ear-mites or some other irritant.

Excessive thirst. Do not deny your cat water in these circumstances. If a cat gives the appearance of being thirsty, yet does not drink, this may be a sign of Feline Infectious Enteritis, which is serious.

Eye problems. If your cat is feeling 'low', its eyes may shield themselves with an extra eyelid. This may look alarming, but it is a useful warning sign, and will disappear as soon as the cat's ailment has been identified and successfully treated. If the eye is simply watering, or has a discharge, bathe it with dilute salt water, then take the cat to the vet.

Salivation. Cats sometimes salivate when they are expecting a tasty meal, but persistent salivation may indicate poisoning or flu, in which case it is important to see a vet. If your cat has 'bad breath', this may indicate decaying teeth, but it could result from kidney problems.

Coughing. Your cat may have something caught in its throat; it may have an allergy; or it may have flu. Persistent sneezing is likely to indicate flu, which is both infectious and dangerous among cats.

The clean cat

Many people choose to keep a cat because they believe it to be fastidious and clean in its personal toilet habits. This is certainly true of most cats, and the reason may be, as some experts have suggested, that they simply cannot stand any sort of mess around them.

They do not even like using a dirty litter tray, but will do so if they are unable to go out.

One can only guess at the reasons for the cat's behaviour in burying its own body waste but, as with most animals, the pattern is predetermined, and seldom varies from that of its ancestors, no matter how remote they are in the evolutionary chain.

The most obvious reason for the burying ritual is that the cat wishes to avoid leaving telltale signs that could lead an enemy to the nest. However, just to underline how perverse they can be, cats sometimes do it not for concealment, but for territorial marking.

The cat may also feel that in burying its excrement, it is reducing the chances of itself, its young, or its extended family, being reinfested by its own internal parasites. However, it may be deluding itself, for with increased urbanization and domestication, in addition to smaller toilet areas, the opposite could happen. The worms and their eggs become concentrated in a very small area and the cat is, in fact, cultivating a garden of parasites injurious to itself.

Another reason for the behaviour may be the certain amount of satisfaction involved in following a ritual, no matter what the purpose, and there may be frustration if it is curtailed or hurried in any way.

While the mother cat is nursing her young, she keeps the nest clean by the simple expedient of swallowing everything they leave behind. However repugnant to humans, this practice must have evolved because no rational alternative was available.

Cats will be cats, whether in town or country. The urban animal will dig into a litter tray, made of plastic or steel, in exactly the same manner as its country cousins dig into loam. Like a programmed computer, it cannot stop itself.

Covering up

Cats cannot prevent themselves from covering up. The pattern is pre-determined and hardly varies from that of its ancient ancestors in the evolutionary chain. The ritual could be performed to avoid leaving signs which might lead an enemy to the home territory, or it may be a form of territorial marking. The mother cat may also feel that in burying its waste, it is protecting its young

Cats are very private creatures and they will take a great deal of time and trouble to find the right spot for their toilet activities. This is usually in some secluded part of the garden, well away from inquisitive human eyes.

Moving your cat

There may be many reasons for moving your cat from one place to another. You may move house or your cat may become ill and need veterinary treatment. Cats love their homes. There is simply no way to explain to a cat that moving house is either unavoidable or beneficial. They simply find it upsetting. A sick cat will find even the shortest journey doubly disconcerting.

There is only one safe way to move a cat. Place it in a secure basket. Neither conversation nor tranquillizers are effective substitutes. Admittedly there are some cats that travel the world perched like parrots on the shoulders of their owners. But the overwhelming majority leap off at the first opportunity and are never seen again.

Cat baskets should be light enough to carry easily. They should close and lock securely. They should contain no spaces through which a cat can push its limbs or its head. The inside surfaces should be smooth. Frightened or angry cats will hurl themselves about. The basket should be large enough to easily contain the adult cat but not so large that it gets thrown from corner to corner. The materials from which it is made should be durable, light and washable.

Some owners insist that, before a long journey, their veterinary surgeon should prescribe tranquillizers. They can be effective but some cats react adversely. The poor creature may go berserk, or huddle in a corner in its basket, grooming profusely, and shivering uncontrollably. First test your cat's reaction with a minimal dosage some time before your journey. If your cat reacts badly you will just have to settle for a noisy trip.

Some perfectly healthy cats actually enjoy travelling, and owners sometimes risk having a cat loose in a car during the journey. This is not advisable. On the other hand, a cat that enjoys train rides might like the comfort of its owner's lap or the view from the window.

If you move house, keep your cat strictly confined until you and the cat have settled into the new home. A good alternative is to board animals until your move is complete. In this way, the chances of losing a cat are minimised and the artificial break may help it to accept the transition.

When you first allow your pet outdoors for its first exploratory wander, make sure it is hungry. Don't give it its usual breakfast, and then about fifteen minutes before dinner time let it out. It will go through a cursory reconnaissance, but its empty stomach will ensure that it responds to your call for dinner.

Neighbouring cats and dogs have to be gradually lulled into acceptance of the new resident. To ease the transition into the new neighbourhood, some cat experts suggest that, instead of disposing of the contents of the litter tray in the usual method, while the cat is settling into its new home distribute it around the edges of the property. After a couple of weeks of this sort of 'odour inoculation', the neighbourhood animals might accept that a stranger has moved in and intends to stay.

Despite all your care and love there are some cats which against all the odds will set off to find their old and familiar homes. Those who succeed make news. Scientists say they are using a behavioural mechanism or aid called *orientation*. But even in migratory birds, whose very survival depends on that ability, the mechanism is still imperfectly understood. Cats are not considered to have that facility. Sadly, most of the cats which set off to find their old homes are never seen again, no matter how strong their motivation.

Steerage . . .

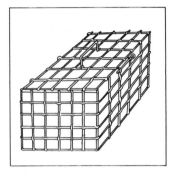

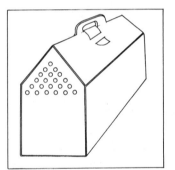

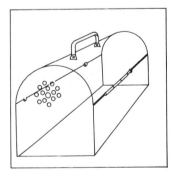

. . . or first class

Some vets say that there are no difficult cats – only difficult owners. Some are even classified as dangerous. There are those owners who try to impress with their own veterinary knowledge, and those who will not accept the assistance of the nurse but have to see the boss. Secure your cat in a basket to take it to the animal hospital. Don't let it out once you are in the waiting room. It is not a Noah's Ark. Every animal in the place will be a stranger and they will all be ill or injured, as well as frightened of their surroundings. They need the comfort of conversation as well as the expertise of the surgeon.

The best sort of basket is made of wire or metal coated with fibreglass or plastic. A basket like this will last for many years, can be easily cleaned, and permits the owner to keep a close watch on the cat. Those made of woven reeds are good provided the ends of the reeds face outwards. They can be sterilized with bleach – steam or hot water appreciably shorten their usefulness. Hardboard carriers tend to fall apart in a shower of rain. A determined cat can push its way through the flimsy sides. The perspex box gives clear vision to the cat and its owner. The interior is smooth for cleaning and has no sharp edges to harm the cat.

All baskets should open at the top. It is easy to place a cat into a basket with an open lid and it can be just as easily removed. Those that open at either end or side necessitate a pushing contest to get the cat in, and a battle to get it out.

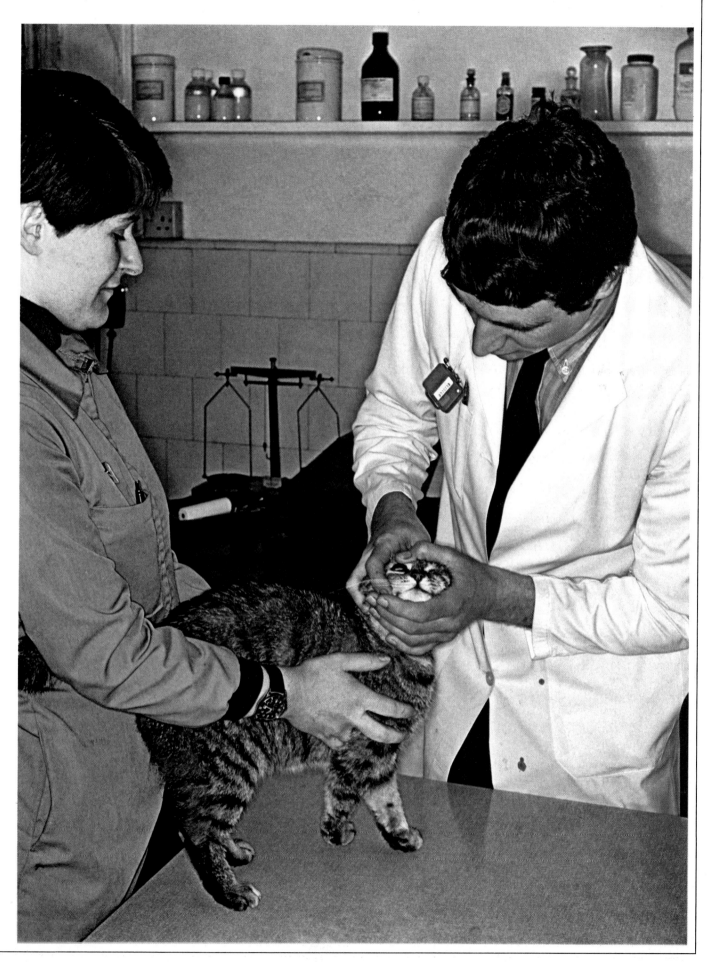

Lifting and carrying

A cat must be held safely and securely and never hurt during the process. If possible, handle it so that it does not hurt you either, which is sometimes easier said than done.

Young kittens may be held and carried exactly as the mother does it. Grip the scruff of the neck and get on with it. The kitten will not mind, for its neck area is strong and its body weight light. As it puts on weight, this method will cause pain, and the growing cat will react accordingly.

The grip on the neck remains a good way of initially securing the animal, supporting the body with a firm, but gentle grasp of abdomen or hind-quarters. If you are moving a kitten a fair distance, snuggle it next to you, while retaining a grip on the neck. Point its face and front legs away from you. If held securely against your body, the kitten should feel safe enough to relax.

Although most children have little knowledge of how to handle a cat, they seem able instantly to communicate their feelings to the animal. Often, even the most disagreeable feline will respond by leaping into the child's arms and cuddling into the most comfortable position, purring all the while.

The cat may even join in a game in which it is dressed in an assortment of bonnets and socks, or even wrapped in bandages borrowed from the first aid box.

Some cats are almost impossible to handle by the scruff of the neck. The most obvious are full toms in peak physical condition. They have no loose skin to speak of, and their necks and shoulders are muscled like a bull terrier. Few men have enough strength in their fingers to hold an unwilling tom in one hand.

Contrary to popular belief, many of these powerful cats are among the gentlest of creatures, if they are not riled. Try a little soothing conversation with the animal before going any further.

In common with many elderly cats, some younger ones suffer from arthritis of the spine and neck and simply cannot bear pressure on these joints. The neck area is a typical site of heavy flea-bite infection, or abcesses. One can hardly blame the cat if it responds adversely to squeezing of a painful area. The sensible solution is to look before you reach.

How does one hold and carry a cat without applying neck pressure? A gentle cat, that knows and trusts you, will often allow you to scoop it up with a hand under its chest and the other around its hind-quarters or abdomen. Others may have to be picked up with one hand gripping the fore-legs and the other the abdomen – the method used by judges during competitions at shows.

Correct and comfortable

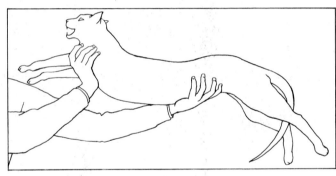

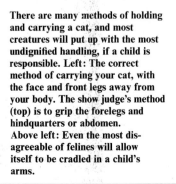

There are many methods of holding and carrying a cat, and most creatures will put up with the most undignified handling, if a child is responsible. Left: The correct method of carrying your cat, with the face and front legs away from your body. The show judge's method (top) is to grip the forelegs and hindquarters or abdomen.
Above left: Even the most disagreeable of felines will allow itself to be cradled in a child's arms.

Above right: A kitten can safely be held by the scruff of its neck alone, for its neck is strong, and its body weight relatively light. The child pictured on the right is not holding her pet in the correct manner, but it does not seem to mind. However, the same animal's immediate response to similar handling from an adult, even a member of its own human family, would be much more dramatic and much more violent.

Holding operation

Many animals are on the move almost from birth, and the new-born antelope, for example, can outrun a fully-grown wolf.

By contrast, the kitten is a helpless creature, totally reliant on its mother which is, in turn reliant on the security of the nest she has built. If that security is threatened, she must move the kittens.

There is only one efficient way of doing this, and it is to grab them around the neck. The tail would not afford a hold, and neither would a limb. In either case, the kitten would bounce around and risk serious injury.

The neck and shoulder area has the great advantage of being near to the kitten's centre of gravity. The hunting cat carries its prey in exactly the same way, for the same reasons.

Those of us who have had to carry a cat, have discovered that they are less able to respond with a violent argument if their neck and shoulders are under strict control.

The mother may feel obliged to move for many reasons. A cat out in the wild may have to cope with such things as combine harvesters, which can flatten a thousand acres before breakfast. Few rural nests can be considered inviolate. Unless the nursing queen responds immediately to the sound of advancing machinery, the result is death.

In the domestic situation, the movement of kittens is seldom so pressing. The main reason for the switch may be neurosis, brought about by well-meaning, but unwelcome members of the human family.

Yet another reason may be cleanliness. As the kittens grow in size, so does the mess they make, and the fastidious queen may not be able to cope. Like the loaded oil sheik who changes his Cadillac because the ashtray is full, she moves house.

Unless the mother has had her kittens by the back door, she must make a move when the youngsters begin to explore. The kittens will need more space than the warm cupboard will allow to exercise their growing physical strength, and by then the mother will want to show them off to you.

Mother knows best

The domestic cat may feel obliged to move her litter for many reasons, although the most likely is human interference. She will then find a new nest and move the kittens one by one. Carrying them by the neck is not instinctive. It is, however, the most convenient method and mother soon learns what is best.

Play or practice?

Play is basically a preparation for the realities of life, but how can we tell when an animal is playing? First, it does not seem to have any purpose and, second, the animal expends just that bit more energy than is really necessary. Simply, play is almost always characterized by exuberance, if not joy.

Watch closely and you will note that your cat will stop playing the moment a real problem or need arises. The kitten chasing a ball will suddenly forget all about it in order to deal with a flea. A litter of kittens engaged in mock combat will break off in mid-sprawl if they hear the clink of dinner bowls. However, a kitten taking part in a serious activity, like eating or stalking prey, will not allow play to intrude.

Some play has an obvious purpose or lesson. The cat playing with a toy mouse is preparing to deal with the real thing. Kittens which play-fight with each other or their mother are learning the art of feline defence and offence. Many an orphan reared by humans, without the company of a litter, has grown up into a timid and bewildered adult.

What of the many play activities for which there appears to be no reason? Dark, enclosed spaces hold an endless fascination. Tubes, tunnels, chimneys, washing machines, fridges, boxes or bags are just irresistible.

Some students of animal behaviour make the Freudian suggestion that it is simply a way of returning to the security of the nest. Some say they merely want a quiet place to sleep.

A rather far-fetched theory says that contact with the sides of the box or bag gratifies the cat's basically sensual nature. The same suggestion is made to explain why a cat will wind itself up in a ball of wool.

More likely, the cat finds in the many twisted strands endless objects to chase and countless enemies to vanquish.

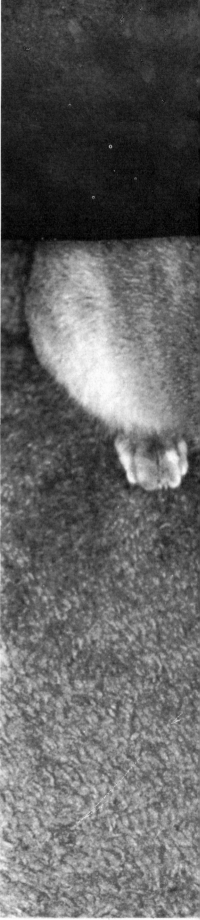

Almost anything has play value for the imaginative and fun-loving cat – silver paper with its fascinating glitter, or even a simple piece of fabric. However, what really grabs the interest of the average self-respecting feline is something like a tube, box, bag, chimney or even an object as unlikely as an open washing machine (below). There is no definite reason for this behaviour.

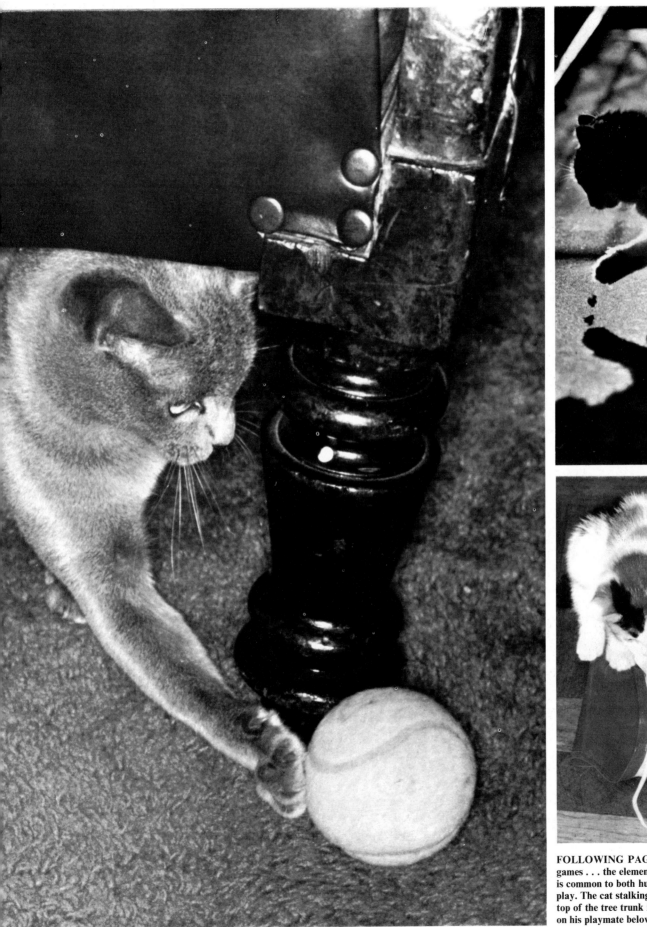

FOLLOWING PAGE: Predatory games . . . the element of surprise is common to both hunting and play. The cat stalking along the top of the tree trunk is practising on his playmate below.

Just for fun

One wonders how different the tests of animal behaviour would be if the academics who dominate the field were not the efficient and single-minded Germans and Americans. It is a safe bet that if the Italians were masters of this art, words like joy, pleasure, love, vitality and adventure, would appear far more frequently.

It might even be accepted that many activities have absolutely no measurable purpose in terms of survival.

You may think about it as long and hard as you like, and then feed your conclusions into a computer, but it is doubtful whether any scientific reason will be forthcoming as to why your cat enjoys playing with your typewriter.

An uncommonly high proportion of artistic people, such as painters, sculptors, writers, musicians or actors, are cat devotees. This fascination certainly cannot be attributed to either stupidity or boredom. Perhaps it should be accepted that artistic sensibilities perceive motivations of cat behaviour that the more prosaic scientist has not yet managed to measure.

However, even scientists will admit that some creatures, such as otters, dolphins and several crows, play for no other reason than pleasure.

The mother cat encourages play, and she supervises every minute of it. The kittens observe her carefully, and learn her signals and attitudes. Expressions of threat are 'taught' by the mother, and then the kittens try them out on her.

The importance of this interplay is not as vital as it may seem, for orphaned kittens, reared alone, do not seem to have any difficulty in mastering the art of being threatening.

The most common forms of play are those which resemble fighting, escaping and hunting, but the distinction between fun and reality is striking.

A simple comparison will illustrate this point. A pack of hounds leaps the fence and goes for the cat. The terrified animal literally has to run for its life, and it will not stop until it is well and truly out of danger.

If, instead of the hounds, a friendly neighbourhood dog had jumped over the fence, the chase would have been a game greatly enjoyed by both participants. In fact, after a few yards, the escaper would turn and chase the attacker. Throughout the game, claws would remain sheathed and the teeth would not be bared.

Another major difference between the two situations is that the cat which is merely playing a game will instantly change from one posture to another. However, if it has recently escaped from real danger or threat to life, it will remain fully vibrant for several minutes.

In other words, a genuine activity has a clearly defined beginning and end. As it is when children play tag with each other, the game can start and end in the middle.

When survival is a genuine consideration, energy is conserved; in play there are no such limits.

Just as it is with children, play is a vital part of a cat's growing-up process. Real life will always come first, for the cat will instantly forget about its game with a ball as soon as it hears the clink of its dinner bowl coming from the kitchen. Furthermore, no cat will permit play to interrupt a serious activity in which it is involved, such as eating its food or stalking prey.

The participant

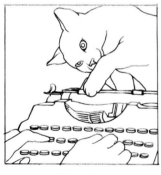

Cats are ready to join in anything that their favourite human is doing. This may be merely watching the television, working at the sewing machine, playing the piano, or using a typewriter. Cats do not mind, for they are only too happy to 'help', whatever is going on.

A healthy interest

Far from killing the cat, curiosity more often saves its life. Strange objects and situations will always be thoroughly examined from a safe distance before closer inspection is felt justified. Final acceptance is a long process. The cat can also quite often be refreshing its own memory, by checking its own domain. If a familiar object has been moved to a new area, it will be subjected to the same detailed scrutiny as a newly-arrived piece. Exploratory behaviour is measurably increased in animals which have constant involvement with humans. Like children in a secure family, they are less afraid to venture into the unknown. Cats from secure homes also tend to live longer than those whose every aware moment is a struggle.

The ardent hunter

It is easy to distinguish between the cat resting on the lawn and the one paying very special attention to something in the undergrowth. The whole attitude of quivering tension and utter concentration makes it clear that a drama is about to be enacted.

Occasionally a bird or a rodent will fall into the cat's capable jaws, but in the main it must painstakingly stalk its prey before it pounces.

The manner in which it does this, and the time it must spend on it varies considerably, depending on where the cat lives. The majority of cats live in well-tended suburbia, where the rodent population is rigidly controlled by man, and the birds enthusiastically protected.

In this sort of affluent society, the cat may have to spend hours in search of its prey. Sometimes it may spend the best part of a day patiently waiting by a mousehole, or crouched in a tree. The young cat is more likely to persevere, the older one becomes bored. This hunting cat easily becomes a sleeping cat.

The farm cat, or one which operates around factories and the garbage bins of the modern city centre is more fortunate. For it, the hours of crouching are likely to be reduced to mere minutes.

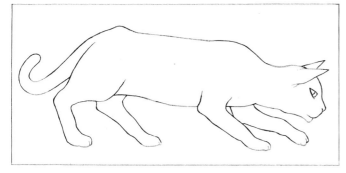

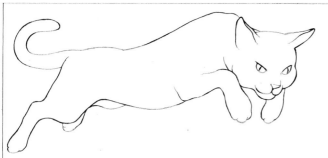

Crouching to kill

From a view to a kill. Left: the cat spies its prey and goes into a crouch, waiting for the right moment before beginning the slink-run. Head, body and tail seem to glide along the ground as the cat makes ready for the pounce (below left). Toes dig in, heels rise and the hind legs move back. The tip of the tail reflects this contained energy by twitching. The whole posture is similar to that of a sprinter waiting for the gun. Finally, it pounces, but always keeps at least two feet firmly planted on the ground – just in case it should need a sudden means of retreat. Above: the small drama is complete as the cat lands on its unfortunate prey, seizing it with mouth and forelegs. Without assistance from a human, the prey is doomed to die.

The cat of prey

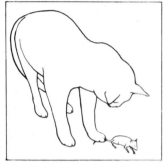

The playful predator

Why do cats torture their prey before they kill it? It does not occur to the cat that it is inflicting pain, for torture is not something it understands. The cat is simply releasing its pent-up hunting energies. The most common way of 'playing' with a crippled victim is the shake. It is also commonly observed in dogs, and most other carnivorous predator. The reason for shaking the victim is obvious: a violent shake or two will disorient any captured creature. It will upset the balancing mechanism of the ears. For vital seconds it will be unable to stand upright – and will certainly not be able to fight back or run away. Some cats hold down the prey with one or both forefeet, while they worry the rest. This behaviour may be related to the evolutionary past and the killing of snakes.
PREVIOUS PAGE: Cats will sometimes try to take on more than they can manage. This cat has attempted a mass attack – and all the birds survived the ambitious onslaught.

Much has been written and said about cats – such as the Scottish Wild Cat – which are supposed to go fishing 'for a living', but it is impossible to authenticate any of these accounts. Of course, there may be some streams so full of fish at certain times of the year that the well-directed hand or paw can ensure a hearty meal. However, there can be very few places with enough catches to make year-round survival possible. Cat, it seems, cannot live by fish alone.

They are inordinately fond of fish, but it is obvious that this undeniable taste has been fostered by man. Cats tend to congregate at the docks, around fish processing plants, and as close to restaurants as man will allow; but this can surely be classified as scavenging.

Similarly, it must stretch the very bounds of credibility to suggest that any cat can satisfy more than a tiny fraction of its nutritional needs, let alone its appetite, by hooking your pet goldfish.

Admittedly, the occasional fishbowl or ornamental pond is partially depleted by the family cat, but hunger is prob-

ably not the likely explanation. The curious cat will spend hours watching a moving object, and from time to time may dip in a paw. Unless it is very lucky, this action will prove futile. After a few days of this, even the most dim-witted cat will start to realise that the likely profit margin is too small to justify the effort. From then on, the goldfish, the canary, or the hamster, assume the same importance to the cat as the television set.

Other moving creatures, like flies or invading mice, never lose their fascination for the cat. This is simply because the cat's long hours of watching and waiting are quite likely to be rewarded. Such rewards, no matter how few and far between, are the stimulus for renewed effort. It is not too difficult or unreasonable to draw a parallel with those gamblers who spend hours feeding coins into fruit machines.

The domestic cat has little control over its own environment. It cannot even control its own population and distribution. It may have limited control over the rodent population in its immediate area, but basically its predatory behaviour has been tempered by its proximity to man.

That is to say that the things we see might not be the same if they were done in public. It is quite likely that when cats are out of human sight, they do not 'play' with their crippled prey for hour after tortured hour.

We have all seen cats stalk and worry a mouse. Only when the poor creature is actually exhausted into unconsciousness does the cat bother to eat it. Some owners are quite dismayed to discover their cat under the sofa gnawing a creature that it must have killed a day or more previously. This is because the cat, returning with its prey, has been distracted by some other household activity, such as the preparation of the family's evening meal. It has then hidden its rodent snack until it can be relished at leisure, long after the kitchen scraps are a memory.

Why does the cat invariably run off with its trophy? First, the cat probably seeks privacy and security; second, the phenomenon of food envy. The cat wishes to prevent a rival from eating the catch, even if it does not want to do so itself.

An instinct shared

Cats are natural hunters. They are typical carnivores with jaws and teeth intended for eating flesh. Most domestic cats prefer to be lonely hunters, although it is not uncommon for them to hunt in pairs or even in groups. Kittens and cats play together, and play is a safe way of learning to hunt. It is observed that a mother cat will bring her young an injured bird or rodent, even though the kitten has no need of food. The reason appears obvious: she is teaching the kitten to pounce and kill. Later, when the offspring is more advanced, she will introduce the refinements of reconnaissance, ambush and the slink-run. The cat relies on stealth and surprise in hunting its prey and, although this killer instinct is regarded as one of its least attractive qualities, cats can be thanked for keeping down the vermin which would otherwise cause far more problems in society. Contrary to popular belief, cats are not great bird-catchers and usually have to make do with the lame and the halt. In this way they are assisting the natural selection of nature.

Sabre-rattling

Animal aggression is a ritual, applying to cats as much as other species. The real name of this awesome game is intimidation.

Take as an example creatures with antlers. They work off anger or establish pecking order by butting each other, and they always meet head on. However frightening it looks, or how heavy the blows that land, it all forms part of a ritual designed to save the species from wiping itself out.

Their skulls may ache a bit afterwards, but both animals can walk away from battle. If the aggression was more than ritual, the stronger would have gone for the flank of the weaker. If the area between ribs and hind leg was penetrated, it would mean certain death.

These animals instinctively refrain from asserting superiority over each in such decisive fashion. Cats behave in the same way. This is how it must be, for those killing their fellows simply do not survive.

Feline signs of aggression are obvious. The hiss, and the well-displayed fangs; the pinpoint pupils; the ears pricked forward, even parallel with the slope of the head; every hair on the body raised.

The sound and the fury

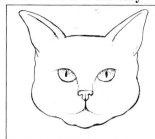

Feline aggression is not a pretty sight, and it is not meant to be. Claws out, spitting, the teeth bared, cats go through a time-honoured, ritual war-game. The real warning sign is the backwards flattening of the ears. Only when they actually go on the attack do the ears flick forward. Deaf cats are among the few that do not issue a warning. They go right in with claws.

Public Enemy Number One

Cats can, and do, get on well with some dogs. Others are unwelcome. In this photograph, a canine interloper has disturbed two cats from their lofty perch. One has immediately jumped for safety. The other is clearly made of sterner stuff and is fighting back. Faced with this resistance, the average dog is likely to retreat.

Everyone knows the old saying, 'They fight like cat and dog'. The dog is, of course, the cat's traditional and time-honoured Public Enemy Number One, but why should this be so?

The main reason why cats are involved in more confrontations with dogs than any other alien species is simply that there are more of them about. Cats are willing to share their homes with dogs, but they are not all fond of canine outsiders.

Generally, cats are not afraid of other animals smaller than themselves. They are relegated to the status of potential meals, or just amusement. Creatures of equal or larger size are potential enemies and are treated as such until proved to be otherwise.

However, the cat certainly does not look upon even the tiniest of miniature dogs as a snack, and the powerful Great Dane would never consider eating a cat.

Obviously, the reasons for their inevitable hostility come down to such important animal matters as security, status and territory.

Many cats fancy their chances in a clash with a dog, and it is easy to see why. Most domesticated dogs are not really equipped to deal with the sheer ferocity and aggression of a hissing, spitting ball of fur, with sharp teeth and claws.

This would, of course, depend on the type of dog involved and, if there is more than one canine opponent, the ever-sensible cat will quickly opt for discretion rather than valour.

Most dogs in packs, particularly hounds, are potentially lethal, and not only to cats. These anti-social creatures are just as dangerous to other dogs, and even children who get in their way.

Almost as dangerous is any group of greyhounds in training. They too will attack not only cats, but other dogs. Irate owners of savaged pets have made many demands for these racing animals to be muzzled while they are out on the streets.

Alsatians, beagles, and most terriers and corgis, are other canine types which the cat should try to avoid.

Oddly enough, the formidable-looking bulldog is a pushover for the average cat. However, at the bottom of the list of potential threats is the pug, which could hardly be less pugnacious. Its ferocity is entirely vocal, and it will be treated by the cat with disdain and a tail in the air. Because of its size, the pug runs the risk of serious eye injury if it foolishly chooses to stand and fight.

The cat living on the farm, or in the country, has three enemies which should be approached only if an escape has been planned in advance.

These are weaning sows, confined dairy bulls and, most dangerous of all, cornered rats.

A brave face . . .

The drawing shows that even a tiny kitten, faced with an unknown canine, will put up a display of aggression that will warn off even a large dog.

Beware of the cat

Cats have been on the earth for about 50 million years, but they have been domesticated for only 3,000 of them. Therefore, it is not surprising if they have not yet quite accepted humans as their equals.

The alley cats which abound in the cities do not take to human handling at all. Some are difficult, but most are downright dangerous. If they were not, they would hardly be able to survive.

Many well-intentioned people have learned this the hard way. They rescue a kitten or litter, which has become separated from the mother, but no matter how patient or kind the humans are, these cats will rarely respond in anything other than a hostile fashion. The basic character seldom changes. If they come from wild, or semi-wild, stock that is the way they will remain.

Such cats are capable of a limited relationship with a human benefactor. In many cities, kind-hearted people will nightly feed the neighbourhood roamers. The animals will gather at the same spot, and will eat the food provided, but that is as far as it goes.

There will be no leg-rubbing, no stroking, and certainly no mew of gratitude. If the human ever tried to pick one up, the reward would be a clawed hand.

Surely household cats can be equally dangerous if provoked? They certainly can and, indeed, the only difference between them and their wild cousins, is the threshold of tolerance. For example, the tom anxious to get out to find a queen on heat is not likely to be restrained by a human hand.

Cats of whatever sex, whether neutered or otherwise, which have been kept exclusively by a human female, will react unfavourably to handling by a man, and vice versa.

The best way to avoid this situation is clearly to allow kittens as wide a social circle as possible. The more people handle the kittens during their vital formative months, the better.

Handle with care

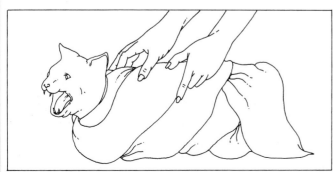

A cat will fight its way out of a situation in which it feels insecure. Even the hand that opens the tins and puts out the food is not immune. If a hand reaches out in an aggressive manner the cat will protect itself. Mere politeness forbids mishandling . . . a wrong approach could mean resorting to the first aid box.

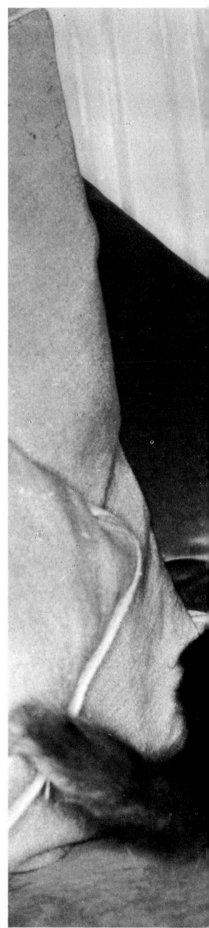

There are many reasons why a cat may go berserk and be aggressive towards family, friends and enemies alike. Obviously distressed, it will tear about the room and be totally unapproachable. There is no mistaking this behaviour with bounding about for sheer pleasure. Something is desperately wrong. If your cat behaves like this in a confined space, do not try to catch it. Protect yourself until it settles down. It will be exhausted. Take a soft garment or towel and wrap the poor creature up, so that it can do no further harm to itself or its human companions. The cat will need veterinary treatment.

The catfight

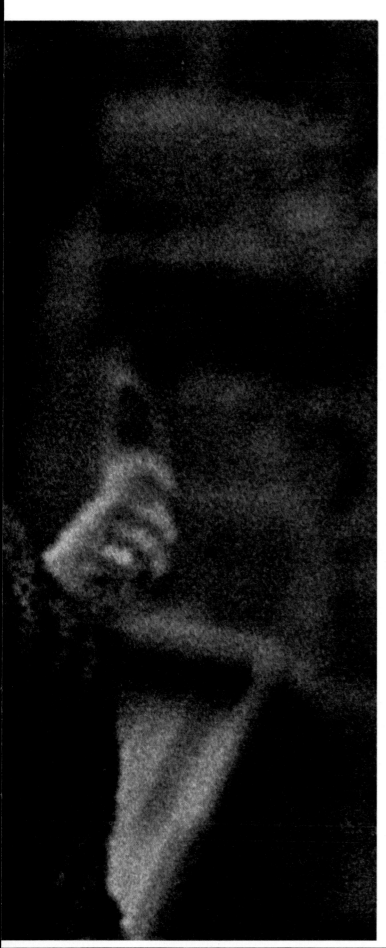

There are four common reasons why cat fights cat – the introduction of a stranger into the home or territory; competition for a queen in season; protection of the young; the release of pent-up energy. Obviously, a new cat in the home, or the house next door, is a threat to established felines, and they will fight for superiority. This does not apply if the stranger is a kitten.

The mating season is something of an endurance test for the tom. He may have to fight off several other males, before reaching the queen, and he must then overcome her own defence of territory.

The queen becomes extremely pugnacious during the final stages of pregnancy, and this continues throughout the nursing period. She will, rightly, fight off intruders, feline or human.

Finally, restricted cats can no more control their occasional aggressive urges than smokers can resist a cigarette. A cat in this crazy mood will attack almost anything in sight.

Territorial imperative

Like most humans, cats regard their home as their castle. This attitude towards territory extends beyond the confines of bricks and mortar. In built-up areas, a cat will regard an area of about one third of a mile as its own; in rural surroundings, a spread of about three miles is not at all unusual.

These distances also depend on sex and the type of cat involved. A female neutered before her first season will feel secure only within the confines of her human family, but she may try to establish territorial rights within a short radius of the home.

On the other hand, the confident and aggressive tom will have a much broader and more ambitious outlook. In all cases, the boundaries of the territory are firmly fixed. Within them, a stranger runs the risk of attack; on the other side of the invisible lines, it will be left alone.

It is, however, impossible to generalize too much, for many highly-domesticated cats will instantly retreat at the approach of an aggressive intruder.

Toms have their own way of marking out territorial claims: they spray a liquid (left), which is highly unpleasant for the humans living nearby, but effective in putting intruders on guard. Right: a typical territorial region around the suburban home. Below: a stand-off between two opposing cats. Often, intruders can be persuaded to leave, simply by staring them out.

Suburban territory

No mans land

The expressive face

The Cheshire Cat, addressing Alice during her wanderings in Wonderland, explained, 'A dog growls when it is angry, and wags its tail when it is pleased. Now, I growl when I'm pleased, and wag my tail when I'm angry. Therefore, I'm mad.'

This endearing creature was surely speaking for itself alone in setting the behaviour of the dog as the standard for normality. Cats are certainly not mad, but they are very expressive creatures, and never more so than when they are angry. There is absolutely no doubt about the feelings of the irate feline. The lips are completely retracted, exposing every tooth in the head, and the ears are flattened so that they blend with the body. Presumably, this is an evolutionary device to protect them from injury, but the only thing apparently needing protection at this point is the would-be enemy.

Other feline emotions and intentions are less obvious, but just as interesting to the cat-lover. Present a cat with something requiring further investigation, and it will advance carefully, ears pricked forward. Each ear may be individually twitched, as if trying to locate the source of possible diversion. 'What have we here?' they seem to be saying.

An unidentified sound is the signal for both ears to be pricked in the direction of the sound. In contrast, the cat which walks about with its ears upright – the normal position – is saying, 'I am on familiar territory. There is no need to pay particular attention to anything.'

In a slightly unfamiliar situation, or one in which there may be reason for suspicion, the ears will be carried in exactly the same fashion. However, they will convey a sense of awareness. No ruler could measure the difference, but it is there, and clearly recognizable to anyone who knows cats.

The cats lips are sensory and sensual organs, used to express pleasure and distaste. When the animal is pleased, the lips will be relaxed, but when it is suspicious, or disapproving, they may be withdrawn into a thin line. The forehead is also expressive, for it will wrinkle in puzzlement, or if the cat is becoming angry.

Most difficult to interpret are the expressions of the eyes. In the first instance, they are shaped by the objects they are trying to see, and the conditions in which they are viewing. The widely-dilated pupil may be seen as an expression of love, but it is more realistic to assume that the animal is merely gathering all the available images and utilizing all possible illumination.

Anyone who has a cat in the home knows that when the animal is feeling completely secure, it will blink or close its eyes. At such times the cat is saying, 'I trust you, and I don't even have to look at you to know that I am completely safe.' The cat's human companions usually find such displays of trust and affection quite irresistible.

At the other end of the scale, cats will keep their eyes closed when they are involved in fights that they cannot win. This is a protective measure, designed to minimize the damage caused by inevitable defeat.

Finally, mention must be made of the common human phenomenon of jutting out the jaw when being defiant. Cats in aggressive or supposedly superior positions will do exactly the same thing. 'Here I am. What are you going to do about it?' they seem to be saying. Only when the fight begins in earnest will the cat tuck in its chin and go on the attack.

The face of the cat presents a mirror image of its inner feelings. Eyes, ears and mouth all give clear indication of the animal's attitude to its immediate surroundings, and its companions. The pictures above, left and below suggest that these cats are impassive, hostile and relaxed, but the cat on the right can surely be experiencing only one feeling . . . that of sheer ecstasy.

The expressive body

The cat's body is almost as expressive as its face as a means of communication, and few parts of the body are more explicit than the tail.

By elevating the tail and bristling the fur, the cat can give most prospective enemies good reason to have second thoughts. No matter how confident the foe, it must be intimidating to find that a one foot high pushover, is suddenly transformed into a menace almost two feet tall.

A dog can do the same thing in a smaller way by raising the hackles on its back, and even the tiny hamster can make itself look frighteningly larger by puffing out its cheeks. However, the cat is in a class of its own.

In addition to the hair on the tail, it can, of course, raise all the hair on its body. It is not clear how much control the cat has over this action, but it certainly has an unnerving effect on the average canine interloper. Tail language is not limited to threat. The same fluffed-up appearance can indicate fear. In this event, the tail is more likely to be held close to the ground.

During ambush and just before the pounce – whether in play or real hunting – the tail is rigid, with the tip gently gyrating. Some would say that it is like an engine warming up for take-off. The angry cat will also clearly indicate displeasure by swishing its tail from side to side. The abject cat may pull its tail between its legs, rather like a dog.

Cats will also use their limbs to communicate. A paw may

be extended in friendship, cupboard love, or in a manner indicating danger. Whether in defence or offence, a paw with the claws out is something which must be treated with respect. A gentle pat with a clawless paw is usually a means of letting you know that the cat is ready for a game, or a stroke. Many an owner has been woken up with the gentle, but unnerving, pat on the face.

The hind limbs obviously are not so communicative. Although they cannot match the rabbit's hind legs as weapons, an adversary can be certain that the cat is in deadly earnest when it brings them into battle.

The way in which a cat walks has a myriad of meanings, most of them quite obvious. For example, the cat which bounds about with all four legs stiffly outstretched, is ready for an energetic game. However, the feline which walks away from a companion, animal or human, in a stiff-legged, deliberately slow fashion, is probably saying, 'You are so far beneath me that you are not worthy of notice at all.'

If this haughty attitude provokes a hostile reaction, the cat will simply become even more infuriating. It will run out of reach and start to groom itself, which clearly says, 'I don't know what the fuss is about. As you can see, I am very busy with my own affairs.'

Everyone has a way of interpreting their cat's behaviour. All cats are different, and we see them all differently. Who is to say how they see us?

Cats say a lot with their bodies. Apart from intimidation or aggression, the arched back can be a sign of pleasure, or a wish to be stroked; so is the far from subtle tap with a paw. The way the cat walks is also a guide to the mood it is in, while the tail sticking straight up with the tip relaxed says, 'It's a great day.'

The cat is an infinitely sensible creature and does not need to be told that going out in the pouring rain is inadvisable and uncomfortable. In sunnier climes, as the midday temperature rises, the wise cat will, of course, decide without any help that a shady spot on the patio is as close to ideal as possible.

The family focus

The mint mystery

Show a cat catnip, and it will show you a change in its behaviour and personality. Even a most reserved animal will shrug off its inhibitions – purring loudly, growling, rolling around and even going to the extreme of leaping into the air. *Nepata cataria*, a species of mint, has this unfailingly ecstatic effect on cats. Even lions are not immune. No one knows why catnip does this, but theorists have suggested that it is an aphrodisiac for cats.

A cat's response to sensation is predetermined. The flickering of the television screen has a great fascination for them, but only if the human family is watching too. Warmth, however, is something they can enjoy without reservation and they can usually be found in front of the fire. Water has the opposite effect, producing a reaction of total distaste.

Cats will sit for hours in front of the television set, but how much they see, appreciate or enjoy is highly debatable. The same cat will continue to sit in the same position, even if the set is switched off, provided its adopted human family are sitting in their customary positions.

Cats definitely can see the images on the television screen, for their eyesight is as good, if not better, than ours. Many learning experiments have shown that they can discriminate between shapes. For example, if they are shown crosses and circles, with one combination representing a reward of food, and another representing shock, they will quickly learn which to choose.

In the same way, cats can discriminate between colours, but probably not to the same degree as humans. There is no way in which a cat's mind can translate the constantly changing shapes it sees on the screen into any sort of meaningful action.

The hearing of cats is possibly more acute than ours. They can hear beyond the human range and into the higher frequencies. This is probably the reason why cats respond more readily to a woman's voice. Incidentally, it is also thought that this sensitivity to higher frequencies enables the cat to hear the voices of rodents as it waits hopefully by the mousehole.

Cats will, of course, prick up their ears in reaction to violent movement on a screen, or to an ear-splitting shriek, or the piteous cries of a baby. Anything that resembles the sights and sounds of their own lives may evoke a reaction. However, by and large, cats nod off in front of the television, and who is to blame them?

As any owner knows, cats are great lovers of their creature comforts. Unerringly, they pick and occupy the best seats in the theatre of life. In many homes the television set is the focus of family relaxation and well-being.

The thoroughly domesticated cat finds much comfort in the company of its family. If that family consists of more than one person, there is almost always a particular individual to whom the cat will devote most attention.

This is particularly true of all queens and neutered toms, for in the normal situation, the whole tom might provide the essential feeling of security. In the human family it is usually the wife who provides the cat's basic necessities, and it is to her that the animal gives its allegience. If she is watching TV, so will the cat; if she decides to run the sewing machine instead, the cat will desert the screen to watch the bobbing needle.

Some dogs will insist on accompanying discordant or high notes with a sympathetic howl. In fact, dogs can hardly restrain themselves from yawning when in the company of people who are obviously bored. Cats do not react in this way, because they lead an essentially solitary life. Neither social howling, nor social yawning plays any part. Like its ancestors, the modern domesticated cat remains silent.

Sight and sound may vary in their effect on the cat, but no stimulus provokes more response than an interesting smell. Everyone knows that felines are greatly attracted to the plant known as catnip, or catmint, though no one is wholly sure why it should have this effect. The smell of food will instantly demand the attention of any self-respecting cat, which is alert to even the most oblique hint of a meal. Most cats dislike rain, but wet weather brings out worms, and therefore birds. For such pleasures, a cat may be prepared to tolerate a soaking.

Familiar faces

Where cats are concerned, familiarity breeds acceptance. As long as they stay together, cats which have been reared in each other's company will remain on reasonably polite terms.

However, like other animals, a cat will react most unfavourably to another which has been absent from home.

The rat, by any standards an intelligent creature, is an excellent example of the extremes to which animals will go. They recognise fellow inhabitants of the home by their odour. Remove the rat from its own area for only an hour, so that its odour changes, and it will be torn to pieces by its neighbours on its return.

With cats, this process of defamiliarization takes several days. They do not go to the extreme of killing the unfortunate returning animal, but they will make its life difficult.

One Siamese breeding establishment reports that it brings its adult queens into the family home on a rota basis for varying periods. This is done because the kittens of

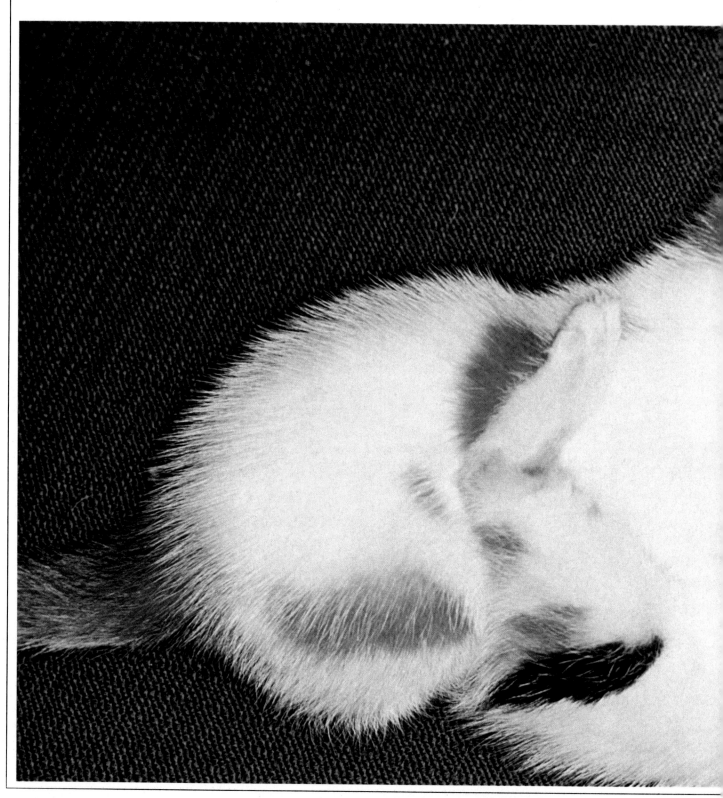

animals which have little contact with family life tend to remain timid and suspicious of people.

The two champion queens, Bella and Tiger Lily, were sisters which had been reared together, and had their first litters together in a communal kennel. The kittens suckled both mothers without discrimination.

A few months later, Bella was transferred into the family home, with Tiger Lily following six weeks later. They acted as if they had never seen each other before, going through the whole ritual of hostility, claws out and hissing. Two weeks passed before they managed to establish a neutrality.

Conversely, anyone with knowledge of animal husbandry will confirm that it is not difficult to introduce an orphan – even one of an unrelated species – to a nursing mother.

The memory of unpleasant experiences lingers longer in a cat's mind. Most vets can tell tales of walking into a house which they have not visited for three or four years to be greeted by the sight of vanishing tails.

A cat's best friend

Why do humans keep pets at all? On the face of it, there seems very little point in bringing an animal into our well-ordered homes. It probably sheds hair, in its early days its personal habits leave a very great deal to be desired, and if hungry it will think absolutely nothing of leaping on to the kitchen table and gobbling everything in sight. All in all, a most anti-social situation.

Man's original reason for welcoming the cat into his cave or hovel was the utterly logical one of wishing to keep the vermin at bay. Some people have suggested, with some conviction, that it would have been easier and more sensible to domesticate the rat.

The cat does have some advantages. For one thing, it is a good deal easier to handle. Rats and cats both bite, but the latter does it less frequently and usually has the courtesy to issue a fair warning.

What other good reasons are there for keeping cats? Perhaps more important than most of us care to admit is the fact that they are soft, warm and cuddly. We also manage to see in them many attractive human expressions and characteristics.

This, of course, extends far beyond cats. Although most self-respecting bears can remove a human head with a casual swipe of the paw, we persist in regarding them as big, lovable dolls. The market for toys resembling snakes, hyenas and frogs is perhaps more limited. We like cats because they look attractive, because they can be well-behaved pets if they are looked after, and because they are responsive to humans. As any successful participant can verify, it takes two to waltz, and it is not too sentimental to suggest that the cat is more than ready to meet us halfway.

The cat, as anyone who keeps one will know, is an eminently sensible and practical creature. This is well proven by its evolutionary history and its successful adaptation to all sorts of environments. Other creatures have been equally successful, and are as intelligent, but they have not accepted man's hospitality with as much readiness as the cat. Perhaps the reason comes down to sheer laziness.

Why struggle for the essentials of life, and indeed the luxuries, if someone else is willing to provide them? The human willingly gives shelter from the elements, protection from enemies, food, warmth and a degree of comfort. In return the cat provides the occasional spot of rodent control, although in most modern homes even that is not necessary.

We tend to take ratting and mousing for granted, but many owners are all too aware that some pampered feline friends will flee in terror from anything larger than a ping pong ball.

For the many lonely people who keep them, cats fill the emotionally empty spaces in a unique fashion. No doubt a psychiatrist would make much of it, but in this important area the cat seems to give every bit as much as it receives.

All my loving . . .

How do you politely introduce yourself? People shake hands, some dogs offer a paw. The really well-educated feline gently slides along human legs, maybe adding a gentle, purring vocal reminder of its arrival. Just in case you need a further message, a flaglike tail is raised in greeting. The message is clear and unmistakable – the cat wants to be introduced.

It takes a stern human to resist the attempts of the rubbing and licking cat to get itself picked up and petted. Another ploy is to lie on its back and look helpless. Once it has achieved its aim, the cat will continue to make contact. Sniffing a human's mouth or nose is a means of recognition and of saying 'hello'.

Man is always likely to take the poetry and romance out of anything. Some scientists insist that the reason cats lick human hands is to restore their deprived bodies with salt. Cat-lovers and less physiologically-minded students of behaviour believe that licking the hand is a mere preliminary to the much more obvious bond behaviours of mouth-to-mouth contact.

The lap is the cat's ultimate target. Some oriental types will not even give the human the option, but will simply leap on to a lap and settle down. The reason cats like lying on human laps is that they are comfortable and secure, and there can be few better reasons for doing anything.

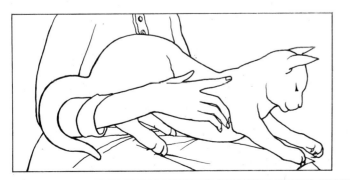

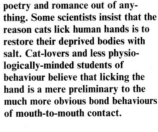

All cat-lovers have been on the receiving end of a gentle, but persistent pounding by paws. This is sometimes assisted, most uncomfortably, by claws, and always with the full strength of the cat's forelegs. This is the normal behaviour of the nursing kitten and it lingers on. Kittens deprived of that sort of behaviour in their early lives are much more likely to inflict it on their owners.

Familiarity breeds content

There are countless homes happily shared by cats and dogs. Familiarity, it would seem, breeds content.

What does this mutual recognition really mean? Simply that all the inhabitants of the home territory, whether they be furred, feathered or hairy, have learned to live with each other. That relationship is not limited to a passive acceptance of the presence of another species. Many creatures actually find comfort and reassurance, if not pleasure, in the company of others.

It is known, for example, that some primates and herbivores share areas to their mutual advantage. Similarly, many birds and earthbound animals share warning signals.

The relationship between the domesticated dog in the home might appear to be less selfish. Scientists hesitate to define any animal relationship as love, but they will admit altruism into the vocabulary of animal behaviour.

Such unselfishness is practised in the animal kingdom every day, and can be seen in such differing species as bee and man. The reason for the worker bee's diligence is obvious: without it, the hive could not continue to survive for very long.

It is not too far-fetched to suggest that cats and dogs share the pleasures and duties of their home territory to their mutual advantage. Neither species will transcend the boundaries of its normal behaviour. We know that when the doorbell rings, the dog barks. We also expect house dogs or guard dogs to exhibit signs of aggression until the stranger either leaves or is welcomed into the house by the human.

Few cats behave in this fashion. Their initial reaction is to retreat and observe. In the home there is rarely any need to be aggressive.

Because of this, the animal's aggressive behaviour is denied a natural outlet so that sometimes it is not only ready to defend its young, but actually appears eager to do so. A naturalist reports that two queens with kittens spotted an Alsatian dog which had innocently wandered into their garden. Without hesitation, they attacked and drove the unfortunate dog away.

In normal circumstances, a cat which has never had a bad experience with a dog will be far more likely to welcome, or at least tolerate, a strange canine. One which has been savaged by a dog, will always be withdrawn and suspicious.

A final word of warning. The cat will almost always accept its own family dog, and vice versa, for each is a recognized part of the territory. Because a dog is willing to accept one particular cat, it does not mean that it will welcome others. In fact, it can cheerfully kill a strange cat venturing onto its home area.

Who needs enemies?

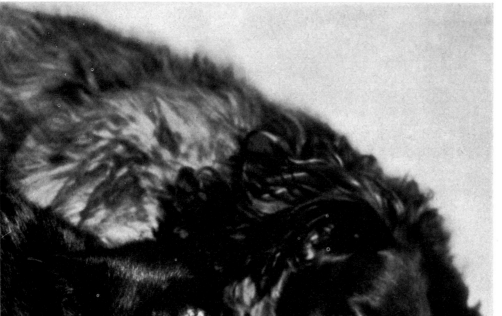

In spite of popular belief that the cat and dog are natural enemies, togetherness is a perfectly normal way of life for these two species — within limits. The dog is a naturally protective animal and will extend its guard duties in the home to all the inhabitants. The cat, in contrast, is a more reserved and solitary creature, but it will accept and, at times, even positively welcome a dog into its home territory.

The social contract

Cats sharing a home usually try to maintain some sort of mutual acceptance, even if this can fall a long way short of affection or respect.

No serious animal behaviourist would ever suggest that cats can recognize their blood relations. This is not to say that such a thing does not ever happen, merely that there is no proof. However, there is evidence that familiarity lessens the normal antagonistic behaviour.

Certainly, cats living communally may not only groom each other, care for each other, and warn of impending danger, but they may even stand together in exceptional circumstances and fight a common foe. This behaviour, which is automatic in dogs, is untypical in cats, for they are not social creatures.

Generally, cats may co-operate only when doing so is patterned on normal behaviour. Cats sleep together as adults, just as they did when they were kittens. The act provides natural warmth, comfort, and a high degree of security. Two, or even three, females will nurse each other's kittens indiscriminately. The maternal drive is so strong that it overcomes the animal's instinctive distrust of strangers. Often the aggressive drive in the nursing female is so powerful that it will co-operate with another to drive strangers away.

However, it does not ever take much for this thin veneer of civilized behaviour to be stripped away, revealing once again the truly wild and untamed nature of the animal.

The experience of the writer and naturalist Graham Dangerfield provides a vivid example of this. He reared a trio of Scottish Wild Cats – two males and a female – from the time they were a few days old. When the female came into season, one male killed the other.

In spite of the fact that the orphaned brothers had lived happily together for many months, familiarity and the fraternal blood bond were completely forgotten from the moment that their innate sexual drive was roused.

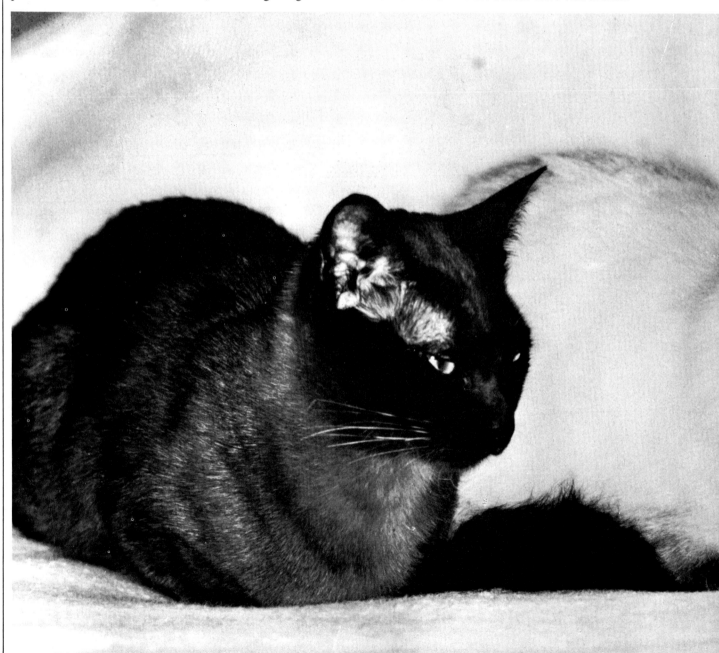

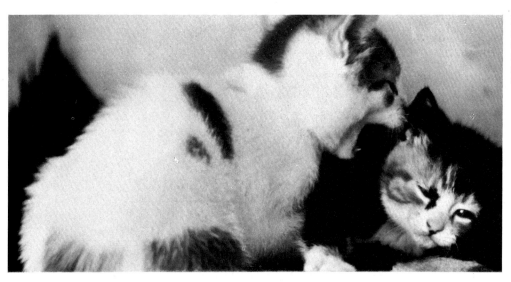

FOLLOWING PAGE: The mother duck in this photograph will brook no nonsense from the cat. Discretion being the better part of valour, the cat is wary of her and her brood of ducklings.

'I said, are you asleep?' There is no doubt that these little brothers are on such good terms that even the shattering of peace and quiet will be allowed to pass with nothing more than a pained expression. This may change when they grow up – particularly if they want the same female.

Friend or foe

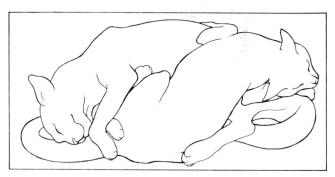

Although they are not social animals, domestic cats can and do co-exist quite amicably. They will care for each other, sleep together and even band together to fight the common foe. The author once kept a pet cat, called Malcolm, which assumed the role of surgery nurse, apparently giving comfort and reassurance to animals coming round from a general anesthetic. It did not seem to matter to Malcolm what kind of animal the patient was, for he gave his time and attention without discrimination to such diverse creatures as cats, dogs, budgies, or even hamsters.

Courting

The queen first comes on heat at about six to eight months, although many have been pregnant at half that age. Generally, the female will be on heat every two to three weeks, and the period during which she actively 'courts' the male is four to six days. If there is no suitable boy friend around, this period can stretch to 10 or 12 days.

If there is a group of eligible males in the neighbourhood, the queen will not necessarily choose the most superior. Experts believe that this is an evolutionary device to ensure that every healthy male has his chance, instead of just the obviously dominant and superior animal. Having chosen, she may then remain faithful. This has been observed to happen from one heat period to another over several years.

What are the obvious signs of heat? A playful kitten will suddenly become more affectionate. She will rub herself incessantly against legs, both human and table. If stroked, she will crouch and raise her tail.

She will often appear quite crazy. If this sounds far-fetched, just ask your vet how many times he has had an emergency call about a kitten, supposedly dying in extreme pain, when the animal has merely been experiencing nothing more than heat. Ask him too about complaints from neighbours about cats being tortured, which also just turn out to be lovelorn.

These signs of heat do not generally appear in the spayed female, which will be a home-loving cat.

The lovesick cat

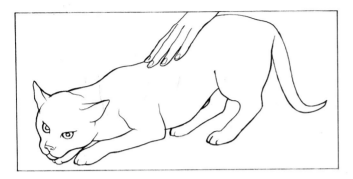

The queen, which is ready for the attentions of a boy friend, will quite often go crazy, crouching, rubbing and rolling. If stroked at this time, she will assume the mating posture (left). If a likely male happens to be in the vicinity, she will call to him (right) and tread with her legs. Courtship is the first of three stages in the mating process, and it usually takes from 10 seconds to five minutes.

Mating

The queen with a mating drive has a far more restricted area of operations than the toms with whom she will eventually mate. In dense urban areas, she may not venture any further than 200 or 300 yards from home.

Even in that limited space there may be danger and she does not like to wander far enough to make the homeward journey any more hazardous.

The tom's attitude is quite different. His actual home territory may be just as limited as the queen's, but such mundane considerations are forgotten at mating time.

Boundaries may be crossed, with all the risk to life and limb that such encroachment involves. Eyes may be scratched, ears may be torn and bodies bruised, but the ardent tom will somehow find a way to the queen.

Once at the mating ground, the toms circle the queen. They keep a wary eye on each other – and the main chance. As one tom moves in, another may attack him. It is squabbles such as these that produce those unmelodious choruses that wake up the entire neighbourhood.

During this noisy process, the queen may roll on her back to attract her would-be lover. The craftiest, quickest, or strongest tom then slants in and mounts the queen. He then immediately moves off. This is to protect himself, either from her reflex attack, or from the impatient rival males. As he dismounts, another tom is ready to take his place.

Although the individual act takes only a few seconds, the mating session can go on all night. As we have said, some queens will allow only one partner, but it is not uncommon for even the most sedate and well-mannered female to accept several dozen before finally running for the haven of home.

The effects of this spectacular promiscuity can be interesting. The feline female is 'superfecund', as the official term describes the condition. What it means is that each of the kittens in her litter can have a different father.

If all the kittens do have individual fathers, they will probably all be differently coloured. This explains the wide variety that can be achieved in one litter.

The battle for the queen

The ardent tom mounts the queen. After all the noisy preliminaries, the sex act itself takes no more than a few seconds. In some cases, the queen may be unwilling to accept more than one lover, but usually she will be the centre of attraction for a group of toms.

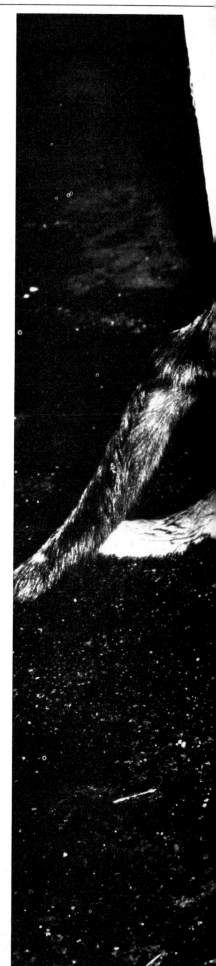

Competing toms will fight (left) for the privilege of being first with the queen on heat. These contests can be protracted, and noisy enough to wake up the neighbours. Below left: the queen rolls on her back to give an added come-on to her would-be lovers. The mounted male (below) grips the queen's neck with his powerful jaws, to prevent her from escaping.

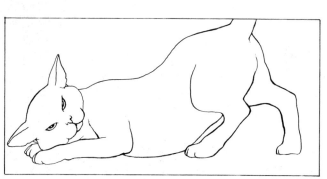

Pregnancy

The queen on heat seeks a mate, and she will not stop calling until she is pregnant. It is then that the loving owner must give extra special care.

For the first two or three weeks, there are no apparent changes, apart from the fact that most queens lose all interest in the opposite sex. Appetite, activities and appearance remain normal.

If you have planned the pregnancy, or if you suspect it, you should now consult your vet about any possible conditions that the queen can transmit to her unborn young. All external parasites, including fleas, lice and earmites, can be eliminated at this stage. It is also thought quite safe to give the queen a booster inoculation against feline infectious enteritis.

How do you know if your cat is a mother-to-be? Apart from a slight swelling of the nipples, there are few signs. Chemical tests and X-rays are not only too expensive but, in the latter case, may be dangerous to the embryo. The best diagnostic procedure for cats is still the loving eye of the owner, allied to the expert attention of the vet.

The pregnant queen is not sick in the mornings, neither does she have a craving for crazy foods. Nature does not allow for such luxuries. All cats need a high amount of protein, preferably from meat, rather than plants. At this time, the queen will require 375 calories per day. While she is nursing, this total rises to 690.

Although her demands will not be outrageous, she will also beg or steal additional morsels. The wise owner provides a variety of minerals, greens and scraps for his cat to nibble at. Anyone who has had a pregnant cat in the house, will know that his pet visits every potential source of food.

Although science has yet to find a way of discovering why, the cat is driven by inner compulsion, beyond its conscious control, to find nourishment not provided by an apparently adequate diet.

During the final 12 days of pregnancy, the queen may not be able to swallow her daily food requirements in a single meal, or even two. She is beginning to feel a bit uncomfortable. You should now offer her nibbles throughout the day, and privacy in which to enjoy them. Also, put down vitamin and mineral supplements in separate saucers and let her choose which one she needs.

The really considerate owner of a pregnant queen will go to the trouble of constructing a heated nest (left) for the mother and her kittens. Right: the queen approaching her time will roll onto her back in order to relieve the pressure and the tightness of her skin.

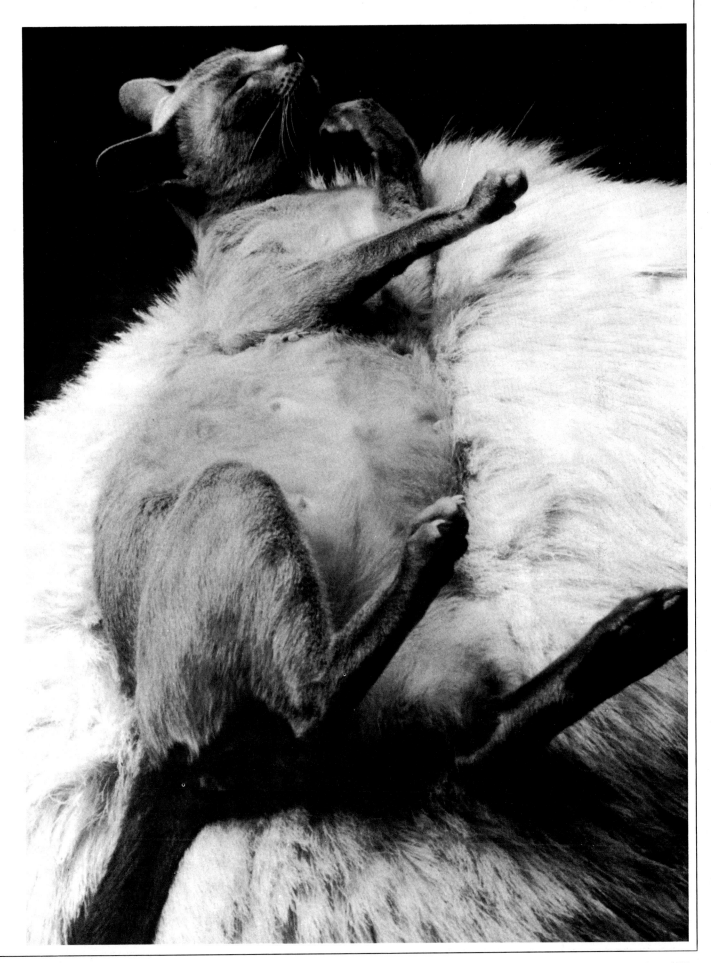

Pre-natal care

Cats are exceptionally healthy creatures. In fact, the goat is the only domesticated animal which can compare. Very few queens abort or miscarry, and the vast majority will sail through pregnancy with no difficulty at all.

Although cats usually make it their business to land on their feet, any sort of fall during pregnancy can have bad effects, causing the animal to miscarry. An attack of flu or enteritis can do the same.

If the miscarriage occurs in the early stages of pregnancy, the cat may show little signs of discomfort. As with humans, a miscarriage in the later stages is a more serious matter and will certainly require a visit from your vet.

Some cats mate successfully, and then come into heat again a few weeks later. In this event, you should either find her another tom, or find out if she is losing her kittens when they are very young. In some cases, a regime of enforced rest and a series of hormone injections are all that is needed to put the animal on the road to normal motherhood.

However, if this happens to a highly-valued, inbred queen, it is likely that the inbreeding has caused some internal abnormality. Such a queen should be spayed.

Many confined feline females astonish their owners by swelling up, making milk, or cuddling up to objects as if they were kittens. These cats may also become morose, neurotic and aggressive and refuse their food. The owners simply cannot understand how their pets managed to get together with a tom.

The explanation is that the unfortunate cat only thinks it is having kittens, a condition known as false or pseudo-pregnancy. If it seems to be making the animal uncomfortable, the vet can prescribe a few hormone tablets. Less serious cases need only time and patience.

The majority of cats sail through pregnancy without difficulty, and survive to pose for a happy family photograph (left) or to get on with the task of weaning the kittens in the privacy and safety of the chosen nest (right).

The happy event

The nesting cat will conduct a thorough inventory of the home area, examining every nook and cranny, before finally settling on the right spot to have her kittens. Her attitude during the search is that of a miser who does not trust the bank.

Essentially, she wishes to have the kittens in seclusion, safe from enemies, and protected from draughts and damp. The chosen place may be something as simple as a cupboard, as maddening as an area under the floorboards, or as inaccessible as a recess under the boiler.

Some cats grudgingly decide that what is good enough for people, is good enough for them, and they have their kittens in bed – your bed.

This cat is usually affectionate and dependent. If it is the first litter, she will usually be frightened, and will need human company and reassurance. Later, she will wish to display her young ones at every opportunity.

The cat that seeks seclusion may resent any sort of handling or interference. She may scratch the hand that tries to help her. In blind fear, she may also not be able to stop herself destroying her own kittens.

Some cats will try to have the best of both worlds. They will select a quiet hideaway, but will welcome the presence of a friendly human. Some will allow this favoured human to stand by, but will show aggression to anyone else who attempts to join in.

Most queens can manage very well on their own. Unhappily, if the inexperienced queen is attended by an

inexperienced owner, the result may be a panic call to the vet.

There are no rigid rules governing the length of labour. Although a litter of six or eight kittens may be delivered in less than two hours, a total period of 12 hours is not unknown.

As the first kitten is about to emerge, the queen changes her posture every few seconds. All her movements and positions appear to be particularly uncomfortable, but they are designed to relax the pelvis to ease the birth of the kitten. She will squat, crouch, scratch the floor of the nest, and may brace her body against any solid surface.

The first action after birth is to wash herself and the kitten. While the queen is busy doing this, she does not seem to have any further contractions.

A proud mother shows off her young ones, and does not seem to mind the photographer intruding at all. However, not all mothers are as obliging. The cat that seeks privacy will not welcome any visitation from humans, or the other family pets. Such a cat may even destroy her own kittens.

Index

Index

PICTURE CREDITS: Robert Morley; Prelim pages. Mike Busselle; *13 (left)*, *36*, *37 (top)*, *48*, *49*, *54*, *58*, 59, 60, 63, 64, 66, 67, 68, 69, 70, 71, 77 *(bottom)*, 80, 83, 84, 85, 92, 95, 96, 97, 98, 99, 100, 101, 104, 105, 106, 108, 109, 112, 118, 121, 124, 125, 127, 128, 132, 133, 134, 135, 138, 140, 141, 145, 151. Trevor Wood; 44 *(left)*, 53, 74, 126 *(centre)*, 130/131, 142/143. Anne Cumbers; 16, 17, 18, 20, 21, 23, 24, 26, 27, 28, 31, 32, 34, 35, 36, 37, 39, 40, 41, 43, 46, 47, 133, 141. Sally Ann Thompson; 42, 44, 45, 77 *(top)* 148, 150, 152, 153. G. Kinns; 13 *(top right)*. R. Estall; 45 *(top left)*, 91, 122, 123, 126 *(bottom)*, 126 *(top)*. Jon Wyand; 73, 90, 186. R. D. Hallman; 19, 89, 107. Roger Daniels; 100, 101. Jan Ruck; 45 *(left)*. Popperfoto 147.